A series of student texts in

CONTEMPORARY BIOLOGY

General Editors:

Professor E. J. W. Barrington, F.R.S.
Professor Arthur J. Willis

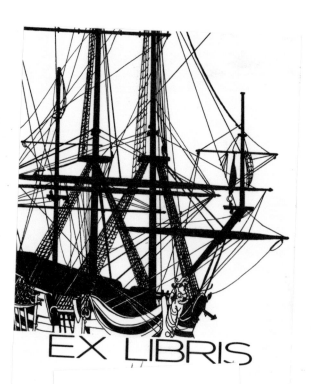

EX LIBRIS

A Biologist's Mathematics

David R. Causton

M.Sc., Ph.D., D.I.C.

Lecturer, Department of Botany & Microbiology,
University College of Wales, Aberystwyth.

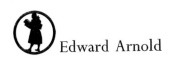 Edward Arnold

© David R. Causton, 1977

First published 1977
by Edward Arnold (Publishers) Limited
41 Bedford Square, London WC1B 3DO

Reprinted 1978

Boards edition ISBN 0 7131 2604 3
Paper edition   ISBN 0 7131 2605 1

Typeset in IBM Press Roman by
Preface Ltd, Salisbury, Wilts
Printed and bound in Great Britain by
J. W. Arrowsmith Ltd., Bristol, England

Preface

Mathematical methodology, in the widest sense, now pervades many aspects of biology; but most biology students are not mathematically minded. This state of affairs is often a headache to teachers of biological sciences in institutions of higher education, and there is no easy way out of the dilemma.

The most widely used mathematical techniques in biology are the statistical methods of analysis of experiments and surveys. These methods have been taught to most postgraduate students of biology for many years, and more recently to undergraduates. However, modern biology requires more of its students than the ability to analyze designed experiments and surveys. The development of any science proceeds from the largely qualitative to the largely quantitative; physics, for example, made the transition about two to three centuries ago, whereas biology is making the transition now.

For this reason an increasing number of biological science schemes of study in universities, colleges, and polytechnics include courses in mathematical ideas and methods, and the main object of this book is to provide a text for such courses. However, rather more is included here than would be deemed to constitute a basic course in mathematics for biologists; the opportunity has been taken to present a wide range of mathematical topics to enable the more advanced student to understand most biological subjects which are treated mathematically. It must be emphasized, however, that the book is primarily meant for first and second year undergraduate courses or equivalent, in keeping with the general aim of books in this series. The path of such students through the book is made clear by means of a plan for a basic scheme of study (page xii) and directions at relevant places in the text. The topics selected for this basic scheme correspond closely with the course given to all first year students of biological sciences here at Aberystwyth.

After an introductory chapter on the role of mathematics in biology, the book considers the fundamental properties of numbers, indices, and logarithms, in Chapter 2. Chapter 3 is also a foundation chapter, dealing with the basis of the geometrical interpretation of many aspects of mathematics relevant to the biologist. A full multi-dimensional approach is adopted; and the topics of location in space, the measurement of distances between points, and linear functions and their geometric representation are covered, examples being given of the biological use of each topic. Chapter 4 then considers non-linear functions and their curves, but in two dimensions only, and also contains a section on some general properties of curves to serve as an introduction to the following chapters on calculus.

The next four chapters deal with the calculus. Chapter 5 describes principles and methods of the differential calculus, while Chapter 6 is concerned with physical interpretation and usage. Chapters 7 and 8 deal with the two aspects of integration: the former chapter introduces the indefinite integral and methods of integration, while the latter deals with the definite integral. Some relatively major portions of these two chapters are best avoided by the beginner, and these are indicated in the text.

Hitherto, the principal type of mathematical function used is the polynomial, although other kinds of function of potential interest to the biologist are introduced in Chapter 4. Exponential and related functions, which are of outstanding importance to biologist and mathematician alike, are described in Chapter 10. Although so important, they are introduced relatively late in the book for three reasons. Firstly, their mathematical properties are not as straightforward as are those of polynomials, and it seems better to use the mathematically simple polynomial functions to illustrate the principles and methods of the calculus. Secondly, students often have difficulty in assimilating the concept of the number e, but the difficulty is minimised if one can discuss e^x as a function whose gradient is always equal to the function itself. Thirdly, it is convenient to define e by means of the exponential series. Accordingly, the preceding chapter, Chapter 9, deals with mathematical series in their own right, as well as serving as a prelude to the exponential series in Chapter 10. The last main section of Chapter 9, which is meant for the more advanced reader, demonstrates how many types of mathematical function can be defined and evaluated by means of infinite series.

Chapter 11 has been excluded from the basic course, not because of great mathematical difficulty, but because the subject matter is specific to plant science. It is included in the book because the material illustrates so well an example of the biological relevance of the calculus. The whole of Chapter 12, on trigonometric functions is considered to be for the more advanced and specialized students, but the earlier parts of Chapter 13 on differential equations should be regarded as part of a basic course in biological mathematics.

Anyone using this as a textbook may thus follow it through in its natural order. There is one possible exception to this. Chapter 14 can be read at any stage, since very little of it depends on material presented earlier in the book, and some of the ideas basic to the chapter now appear in school mathematics syllabuses. It has not been included in the recommended 'first course' because it will be needed at different times in biology courses at different institutions, depending on the applications envisaged.

Mention of applications highlights a particular and ever-present difficulty in teaching mathematics to biologists. Since, for the majority of such students, mathematics is not in itself an interesting subject they are very concerned to see the biological relevance of every mathematical topic discussed. Good biological examples involving *simple* mathematics are very scarce. Biological phenomena are so complex that problems which are not so oversimplified as to be far-fetched require either complicated mathematics, or the application of techniques of probability and statistics. Biological examples are presented whenever possible, but there are sizable portions of the book which are purely on mathematics without any reference to biology. This is inevitable when developing a theme. For example, the convergence of a series and the idea of a limit have hardly any biological connotations, but these ideas lead on to the differential calculus which does have considerable relevance in the life sciences.

Some exercises are provided at the end of each chapter. On the whole, these are to give practice in the methods presented in the chapter, but, where appropriate, exercises involving direct biological situations are presented. Those exercises marked with an * relate to sections of the text which are not included in the basic scheme of study.

Finally, it should be stated that the level of mathematical knowledge assumed of the reader is that of GCE 'O' level, or equivalent, but where possible, topics are developed from first principles.

I should like to express my appreciation to Professor J. Gareth Morris for his constant encouragement at all stages, and for his critical evaluation of the section on the growth of a microbial culture. Dr. Geoffrey M. Clarke read the entire manuscript, corrected my errors, and made a number of other valuable comments and suggestions for which I am very grateful. Any remaining mistakes and obscurities are my own responsibility. Finally, I should like to thank the staff of Edward Arnold for their friendly assistance throughout, and for the excellent clarity of the final production.

Llanrhystyd, Aberystwyth   D.R.C.
1976.

Contents

Plan for a First Course in Mathematics for the Biologist

I

Why Mathematics in Biology?

Probably the most significant event that occurred during the rise of man to pre-eminence, from being merely 'another animal', was the development and use of language. At first, language was only spoken, but it did enable relatively large quantities of information to be communicated from one individual to another. More than that, language also provided a means for controlling and monitoring thought itself. The important point is that language provides concepts which can be manipulated independently of the objects to which they refer, and later it becomes possible to think logically without reference to any particular objects at all. Hence, during the history of man, abstract thought became feasible, and from this beginning arose philosophy.

If we had to select an area of application of language that has been outstandingly successful, we should undoubtedly choose the expression of human emotion. The evidence for this is clear when one considers the achievements of oratory and literature. However, when it comes to conveying scientific information, ordinary language is less successful. This is because it is almost impossible to convey precise meaning, since most words in a language have more than one meaning, even if these meanings differ only marginally. Again, because of these various shades of meaning, a particular word means a slightly different thing to different people. This is fine for the literary use of a language where part of the onus for interpretation lies with the reader or listener, but it is not so good for scientific use, where data and hypotheses must be presented with complete unambiguity.

A long time, perhaps many thousands of years, after the rise of language, a new kind of medium for transferring information from one individual to another was evolving. This was at the time when the first of the ancient civilisations — Egypt and Sumer — were flourishing. These civilizations, besides being the architects and builders of large buildings and other monuments, also developed a calendar. Such achievements required precision: precision in measurement, precision in the subsequent manipulation of measurement, and precision in the transmission of such data to other people. All this could be achieved through another kind of language — the language of mathematics.

The mathematical language is just the opposite to ordinary language in that its elements are precise and unambiguous, or at least should be. Every quantity and symbol used can be accurately defined in terms of earlier quantities and symbols already defined. Thus, mathematics is built up on precision and fact, whereas ordinary languages are, to some extent, based on the variability and imprecision of human feelings and emotions.

THE DEVELOPMENT OF A SCIENCE

The study of any science, viewed historically, consists of two main phases. First it is studied almost exclusively from a qualitative point of view; but after an initial period, quantitative methods come to be used increasingly. One of the main reasons for this type of development, from the qualitative to the quantitative, is that a science begins as an observational study, progresses to an experimental, and finally to a theoretical study. At first, phenomena are observed as they occur in nature; later scientific work consists of performing experiments, drawing inferences from the results, and then trying to formulate general laws. By their nature, some sciences have to make the jump from observation to theory without the intervening benefits of experimentation. Astronomy is an obvious example here, and it is remarkable that observational and theoretical astronomy have proceeded side by side for several millennia.

At the present time, physics is the science which is pre-eminent in the use of mathematics. Many physical phenomena are rather less complex than are those of other sciences, and the subject has progressed through the stages of observation and experiment, and has emerged as a theoretical science. This is not to say that observation and experiment are not still carried out in physics; the main point is that physics has reached the stage in which there exists a substantial body of theory, mathematical in nature, which has its origins in observation and experiment. In physics at the present time, the experimental and theoretical sides of the subject are of equal importance.

The phenomena of chemistry are often more complex than those of physics, and this subject has not progressed quite as far as physics on the theoretical side. Although there have been spectacular advances in theoretical chemistry in recent years, chemistry as a whole is still, at present, somewhat more of an experimental science than is physics.

In biology, the situation is very different. Firstly, biological phenomena are highly complex; secondly, there is almost unlimited scope for pure observation of biological phenomena. Hence it is mainly only in the present century that biology has become an experimental science, whereas experimental work in the physical sciences has been undertaken for several hundred years. As a result, it is only now that 'theoretical biology' is tentatively emerging.

From the above remarks, it would seem that mathematical theory is the ultimate aim in a scientific discipline. This is true; not for its own sake, but because in the last resort the phenomena of nature can only be explained in the precise terms of mathematics. Consider this example from physics.

Observation: a stick immersed in water at an angle to the surface (other than a right-angle) so that part of the stick is out of the water and part submerged; the stick appears bent at the surface of the water.

Experiment: a vessel of water is set up on the laboratory bench, and rays of light are traced through the water for various angles of the incident beam; it is found that at an air—water surface (assume that the vessel is made of very thin glass) the ratio of the sine of the angle of the light beam on the air side of the surface to the sine of the angle of the beam on the water side of the surface is constant, and this constant ratio is called the refractive index.

Theoretical deductions: this experimental result can be used in conjunction with facts gleaned from experiments on other phenomena of light, such as reflection, diffraction, interference, to establish knowledge on the nature of light. For instance, it has been found that the velocity of light in a dense medium is less than in a sparse one. This latter experimental finding coupled with the result of the refraction experiment can be analysed mathematically to show that light travels in a wave form.

For this particular example, there is an obvious relationship between observation, experiment, and theoretical deduction. Such examples can be multiplied many fold. In chemistry, we observe a particular reaction, and we experiment to find out the exact conditions under which the reaction occurs. When we then enquire why this particular reaction occurs and not some other, it is necessary to look to the concepts of physical and theoretical chemistry, both of which are founded on mathematics.

Whether or not all biological phenomena can be explained by the physical sciences, or that ultimately it is found that the property of life is 'something extra', it is already quite evident that the manifestations of 'life' can be explained in terms of the physical sciences, particularly chemistry. Since the physical sciences are based on mathematics, so also, indirectly, are the biological sciences.

In summary, experimental results are usually in a quantitative form, even in biology, and therefore sound theoretical deductions can normally only be made by mathematical analysis. This is why, ultimately, mathematics is indispensable to any science; and so any scientist, whatever his or her speciality, should have an adequate knowledge of mathematics.

BIOLOGY, MATHEMATICS, AND STATISTICS

The mathematical model

From the penultimate paragraph of the previous section, one might infer that the utility of mathematics to the biologist is indirect, arising only after experimental results have been interpreted by the concepts of physical science. This, however, is not so. Mathematics is applied directly to the results of biological observation and experiment in a similar manner to the physical sciences but, because of the complexity of the phenomena, its application is much more difficult.

In the present state of biological knowledge, it is impossible to apply a rigorous mathematical analysis to a biological system, such as may be applied, for example, to an electric circuit. What is done, however, is to construct a **mathematical model** of the phenomenon in which we are interested. Certain **assumptions** about the system have first to be made, and put into mathematical form. These assumptions are based on current knowledge obtained from previous observations and experiments. Next, appropriate **mathematical methods** are applied to the assumptions to achieve an end result which **simulates** the system under study. The simulated result can then be compared with what actually happens. If agreement between the theoretical result and the observed happening is good, then we gain further insight into the process under study; and moreover, we can use the model for predictive purposes. In any science, an ultimate aim is **prediction**. For instance, in an electrical circuit we can predict how the current will change for a given change of resistance, using the mathematical description of the circuit. In a biological system that has been 'described' mathematically by means of a model, predictions of what will happen under certain changes of conditions can also be made. If the results of using the model do not agree with actuality, then one or more of our basic assumptions must be wrong (assuming the absence of mathematical errors!), and so, in a negative sense, our knowledge is still increased. An example of the construction of a very simple mathematical model is given in Chapter 6.

Statistics in biology

There is yet another complication to the would-be user of quantitative methods in biology, and that is **variability**. The phenomenon of variability is not confined to biology, but arises whenever experimental work is undertaken. Even in the physical sciences, repetitions of a single experiment will give slightly different results, e.g. measurement of the refractive index of a substance, or the location of an end point in volumetric analysis. This kind of variability, which is called **experimental variability** or **experimental error**, arises solely because a human being attempts to measure something; the something does not change, but the reactions of the human being during the conduct of the experiment do change.

Experimental variability also occurs in biology, but here it is considerably augmented by the variability inherent in biological material. If we measure the refractive index of a block of glass very carefully, we are safe in asserting that our result *is* the refractive index of this kind of glass, under the conditions of the experiment. On the other hand, if we measure the increase in height of a sunflower plant over one day, we certainly cannot say that this is the growth rate of sunflower plants in general, even under the same experimental conditions. The same plant may have a noticeably different growth rate at an earlier or later stage in its growth; and even if we take two plants which germinated from the same source of seed at the same time, they will almost certainly show different growth rates at any instant, aside from experimental error. So to be able to make any sort of general statement about the growth rate of sunflower plants of a given age and under defined conditions, we have to measure several plants and take an average. This immediately raises the question as to how reliable our result is, and this cannot be answered without employing a branch of mathematics known as **statistics**.

Even when no experiment is involved and one is only trying to summarize observations usefully and build a model from them, a simple mathematical approach may not be very satisfactory because of the variability of biological material. A good illustration is afforded by Example 10.3 on page 186. Read the general description of the situation through, note that a mathematical expression is used to describe the situation, and then carefully read the questions asked, each one of which obviously requires a single numerical answer. Now, without worrying about how the answers were obtained, read the last sentence of each of the three sections, and note that each answer is a precise figure. Bearing in mind that a 'cohort' in this context is a natural stand of similar-aged plants, it is quite obvious that these precise answers are only statements of likely results around which actual results will deviate to a greater or lesser extent. One immediately asks, 'How much deviation can be expected?'. The **deterministic** mathematical model that has been erected to describe the situation in this example cannot answer such a question. If, however, the same model had been set up, but with an added feature — a **probability** structure — then we should be in a position to answer questions like the above. The mathematical model would now be a **stochastic** model; it is much more realistic, and more complex.

Both the mathematics of stochastic models, and of statistical methods for the analysis of experiments, are based upon the same theoretical subject — probability and statistics. It is a branch of applied mathematics in the broad sense, not in the narrow sense that the term 'applied mathematics' is often used to denote applications to physics. Therefore, probability theory and statistical science are based on mathematics, and a good knowledge of the subject is necessary for their study. In this book, we shall not deal with probability and statistics. Our concern will be with such topics in mathematics that are of a general nature, topics that have direct biological relevance, and also those that form a basis for the study of statistical science about which the modern biologist needs to know.

2

Numbers, Indices and Logarithms

Broadly, this chapter will be concerned with numbers and number systems. Numbers, including those defined by symbols (letters), are fundamental to mathematics, and so this chapter should be carefully read and understood even if you find that much of it is revision of material already familiar.

NUMBERS

Imagine a straight line, as in Fig. 2.1, extending indefinitely in each direction. The centre of this line will represent zero, and then at equal intervals on either side of the zero we may mark off points which represent the whole numbers: 1, 2, 3; −1, −2, −3, etc. By convention, positive numbers increase from zero to the right, and negative numbers increase from zero to the left. It is important to note that the symbols +∞ and −∞ do not represent numbers, however large. These symbols may be interpreted in different ways according to their context. Here, they mean that the line extends each way indefinitely.

Fig. 2.1 The real number scale

Real numbers

Any number on the line defined above and shown in Fig. 2.1 is known as a **real number**; in other words, such numbers can be represented physically on a scale. Real numbers are sub-divided further, as follows.

Integer   An integer is a whole number, such as 3, 8, −45, 501.

Rational number   A rational number is one that can be expressed as the quotient (or ratio, hence the name) of two integers. Thus all whole numbers are rational, since each is the quotient of itself and 1. Also, many non-integers are rational, such as 1.5 (=3/2), 2.$\dot{3}$ (=7/3) and −1.$\dot{8}$ (= −51/27). A dot over a figure to the right-hand side of the decimal point indicates that that figure recurs indefinitely.

Irrational number   An irrational number is one that cannot be expressed as the quotient of two integers. Examples are $\sqrt{2}$, $\sqrt{7}$, and π. In general, a number which is neither a terminating nor a recurring decimal is an irrational number.

Thus, a real number may be either rational or irrational; and, if rational, may be either an integer or a fraction.

It may be wondered just why it is necessary to classify the real numbers in this way. The three types of real number that we have just considered would have evolved historically in the same way. The idea of number would have originated in the counting of objects, giving rise to integers. The simple arithmetic processes of addition, subtraction, and multiplication of integers always yield other integers. However, the division of one integer by another does not necessarily give a further integer; so, in order to give meaning to the arithmetic operation of division, another type of number besides the integer has to be visualized. This new kind of number is the rational number; and since whole numbers *can* result by dividing one integer by another, then rational numbers must include integers. When a rational number is a fraction, it either terminates (as in 3/2 = 1.5) or recurs (as in 7/3 = 2.$\dot{3}$). Recurrence is not necessarily confined to one figure: it may be a whole group of figures. Thus

$$6/7 = 0.85714285714285714$$

correct to 17 decimal places, so we can write 6/7 = 0.$\dot{8}5714\dot{2}$ since this group of figures recurs indefinitely. A ratio: al number can be written down accurately either as a vulgar fraction or as a decimal; even 0.$\dot{8}5714\dot{2}$ is an accurate *representation* of 6/7.

Returning to the integers, squaring means multiplying an integer (or any number) by itself, e.g. 3 x 3 = 3^2 = 9. But the opposite process, taking a square root, may give a number which is not an integer, a terminating fraction, or a recurring fraction, i.e. it is not a rational number. To give meaning to the square root, another class of numbers must be designated — the irrational numbers. The main feature of these numbers is that they cannot be accurately written in fraction or decimal form. For instance, $\sqrt{2}$ = 1.414213562 to 9 decimal places; and π = 3.14159 correct to 5 decimal places, or we can write $\pi \fallingdotseq$ 22/7. The sign $\fallingdotseq$ means 'approximately equal to' and it is used here to show that 22/7 *might* be used in place of π in numerical work. How good the approximation is may be ascertained by evaluating 22/7 in decimal form, which is $3.\overset{..}{1}4\overset{.}{2}8\overset{.}{5}\overset{.}{7}$; and it is evident that 22/7 differs from π by approximately 0.00127; a difference of about 0.04% of the quantities under discussion. The decision as to whether a particular approximation of an irrational number is adequate can only be judged by prevailing circumstances, but the only ways in which irrational quantities can accurately be referred to are in the forms $\sqrt{2}$ and π, for example.

Complex numbers

Consider the following quadratic equation:

$$x^2 + 2x + 2 = 0, \qquad (2.1)$$

i.e.

$$(x + 1)^2 + 1 = 0$$

or

$$x + 1 = \pm\sqrt{-1}.$$

Hence

$$x = -1 \pm \sqrt{-1}.$$

Hitherto you have probably been told that $\sqrt{-1}$ does not exist. This is perfectly true if the statement is taken to mean that $\sqrt{-1}$ is not a real number. However, a cursory glance at equation 2.1 does not reveal anything peculiar about it; it is a perfectly normal quadratic equation, but it gives a solution containing a quantity which is not a real number. Now although we can imagine the existence of (say) 2 objects such as loaves of bread, 1/2 a loaf, or even $\sqrt{2}$ of a loaf, the mind boggles at the thought of $\sqrt{-1}$ of a loaf! This quantity cannot be perceived simply because it is not a real number, and it does not exist on our scale of real numbers as shown in Fig. 2.1.

The mathematician does not, however, dismiss equation 2.1 as an impossible type; he invents a new class of numbers and calls them **complex numbers**. The square root of minus one is known as an **imaginary number**, and is always denoted by i. A complex number has the form $a + ib$, where a and b are real numbers, and this is precisely the form that the roots (see page 56) of equation 2.1 take (here $a = -1$ and $b = 1$ or -1).

Although complex numbers are a creation of the mathematician's mind, they are extremely useful in the solution of practical problems. However, the subject is beyond the scope of this book, and we shall restrict ourselves entirely to the realm of real numbers.

The factorial of a positive integer

The factorial of an integer, n, is usually designated as $n!$ (or occasionally as $\lfloor n$), and is defined as the product of n and all preceding integers down to 1; i.e.

$$n! = n(n-1)(n-2)\ldots(2)(1) \qquad (2.2)$$

For example,

$$4! = 4 \times 3 \times 2 \times 1 = 24$$

and

$$6! = 6 \times 5 \times 4 \times 3 \times 2 \times 1 = 720$$

Now although $n!$ represents a real number, an integer in fact, it does not rank with the types of number that we have been discussing above, e.g. rational and irrational numbers. Rational numbers, for example, are a natural sub-class of real numbers, and they have only been given a special name to make it easy to think of them as a sub-class on their own. The factorial of an integer, on the other hand, is an example of mathematical notation. The product represented by the factorial of an integer occurs often in mathematics, and so the mathematician defines this product as we have already done, and gives it a name (factorial) and a symbol (!). Mathematics abounds with specialized notations and many will be introduced throughout this book.

Let us consider two properties of factorials. Firstly

$$n! = n(n-1)! \qquad (2.3)$$

e.g.

$$6! = 6 \times (5 \times 4 \times 3 \times 2 \times 1) = 6 \times 5!$$

This property is useful in cases where we need to evaluate factorials of successive integers; there is no need to multiply right down to 1 each time. For instance, we already know that $6! = 720$; hence

$$7! = 7 \times 6! = 7 \times 720 = 5040$$

$$8! = 8 \times 7! = 8 \times 5040 = 40\,320 \text{ etc.}$$

Evidently $n!$ increases very rapidly as n increases.

The second property of factorials shows that the factorial notation can be used to denote the product of a set of successive integers even if they do not extend down to 1. Consider

$$7 \times 6 \times 5 = \frac{7 \times 6 \times 5 \times 4 \times 3 \times 2 \times 1}{4 \times 3 \times 2 \times 1} = \frac{7!}{4!}$$

The product 4 x 3 x 2 x 1 appears in both the numerator and the denominator of the fraction, and so cancels out. The above example can be generalized. Suppose we have two integers, n and r, and that r lies *between* 1 and n, i.e. $1 < r < n$ (1 is less than r, and r is less than n). Then

$$n(n-1)(n-2)\ldots(r) = \frac{n(n-1)(n-2)\ldots(1)}{(r-1)(r-2)\ldots(1)} = \frac{n!}{(r-1)!} \quad (2.4)$$

Finally, note that factorials of numbers other than positive integers can be defined, but this requires mathematical theory beyond that presented in this book. There is, however, one result of this theory that is important in elementary biomathematics, and that is

$$0! = 1 \quad (2.5)$$

I am afraid that this curious-looking statement will just have to be accepted.

INDICES

A number of the form a^m is defined as the number a raised to the power m; a is usually called the base, and m the index or exponent. For the moment, we assume that m is a positive integer; then a^m means that a is multiplied by itself m times. If we have an expression of the form $a^m \times b^n$ wherein the bases of the two numbers are different, then the expression cannot be simplified further; but if the bases are the same number, we may establish three laws. We shall do this by means of specific examples, and so the table of values below will be useful.

$$
\begin{array}{ll}
2^1 = 2 & 3^1 = 3 \\
2^2 = 4 & 3^2 = 9 \\
2^3 = 8 & 3^3 = 27 \\
2^4 = 16 & 3^4 = 81 \\
2^5 = 32 & 3^5 = 243 \\
2^6 = 64 & 3^6 = 729
\end{array}
$$

Law 1. Multiplication

Example

$$2^2 \times 2^3 = 4 \times 8 = 32 = 2^5 = 2^{(2+3)}$$

Example

$$3^1 \times 3^5 = 3 \times 243 = 729 = 3^6 = 3^{(1+5)}$$

So, generally

$$a^m \cdot a^n = a^{(m+n)} \tag{2.6}$$

Law 2. Division

Example

$$2^5/2^3 = 32/8 = 4 = 2^2 = 2^{(5-3)}$$

Example

$$3^3/3^2 = 27/9 = 3 = 3^1 = 3^{(3-2)}$$

So, generally

$$a^m/a^n = a^{(m-n)} \tag{2.7}$$

Law 3. Powers of indices

Example

$$2^3 \times 2^3 = (2^3)^2 = 8^2 = 64 = 2^6 = 2^{(3 \times 2)}$$

Example

$$3^2 \times 3^2 = (3^2)^2 = 9^2 = 81 = 3^4 = 3^{(2 \times 2)}$$

So, generally

$$(a^m)^n = a^{mn} \tag{2.8}$$

The above three laws should be familiar to you already, and have been derived assuming that m and n are positive integers. We now assume that these laws are valid for all values of m and n; thus indices may be positive or negative integers, zeros, or fractional numbers. The latter may be positive or negative, and rational or irrational. We therefore need to find meanings for the expressions a^0, a^{-m}, and $a^{m/n}$.

Theorem 2.1. The value of a^0 is 1

$$a^0 = a^{(m-m)} = a^m/a^m \text{ (Law 2). But } a^m/a^m = 1.$$

Thus

$$a^0 = 1 \tag{2.9}$$

Theorem 2.2. The value of a^{-m} is the reciprocal of a^m

$$a^m \cdot a^{-m} = a^{\{m+(-m)\}} = a^{(m-m)} = a^0 \text{ (Law 1)}$$

But $a^0 = 1$ (Theorem 2.1); hence $a^m \cdot a^{-m} = 1$. Therefore

$$a^{-m} = 1/a^m \tag{2.10}$$

Theorem 2.3. The value of $a^{m/n}$ is the nth root of a^m

$$(a^{m/n})^n = a^m \text{ (Law 3).}$$

Taking the nth root of both sides, we have

$$a^{m/n} = \sqrt[n]{a^m} \tag{2.11}$$

The argument of Theorem 2.3 may be clarified by a simple example

EXAMPLE 2.1
Evaluate $a^{1/2}$.

$$(a^{1/2})^2 = a^{\{(1/2)\times 2\}} = a^1 = a$$

i.e.

$$(a^{1/2})^2 = a$$

Taking the square root of both sides, we have

$$\sqrt{\{(a^{1/2})^2\}} = \sqrt{a}$$

But the square root of a squared number (i.e. the left-hand side) is the number itself, so

$$a^{1/2} = \sqrt{a}$$

EXAMPLE 2.2
Evaluate $5^{-2} + 3^{2/3} - 2^{-1/2} + 4^{-3/2}$
The expression simplifies to

$$\frac{1}{5^2} + \sqrt[3]{3^2} - \frac{1}{\sqrt{2}} + \frac{1}{\sqrt{4^3}} \quad \text{(Theorem 2.3)}$$

$$= \frac{1}{25} + \sqrt[3]{9} - \frac{1}{\sqrt{2}} + \frac{1}{8}$$

$$\simeq 0.04 + 2.0801 - 0.7071 + 0.125*$$

*This decimal expression is only approximately equal to the above vulgar fraction expression, since $\sqrt{2}$ and $\sqrt[3]{9}$ are irrational.

Therefore

$$5^{-2} + 3^{2/3} - 2^{-1/2} + 4^{-3/2} = 1.538$$

(correct to 3 decimal places).

LOGARITHMS

Previously, your prime concern with logarithms has been as a means to simplify the arithmetic of multiplication and division. Logarithms are, however, very important in mathematics for other reasons, and they are of particular relevance in quantitative biology. Hence we now need to study some aspects of the theory of logarithms.

Definition. The logarithm of a number to a certain base is the power to which the base must be raised to give the number.

Thus if

$$\log_a m = x \qquad\qquad (2.12)$$

this implies that

$$a^x = m \qquad\qquad (2.13)$$

The base is vital. A logarithm cannot be defined unless the base is specified: conversely, the logarithm of a number to one base is different from the logarithm of the same number to another base. E.g. $\log_{10} 2 = 0.3010$ but $\log_2 2 = 1.0000$. It should be noted that the base of a logarithm may be any positive real number; note also that we only define the logarithm of a positive number.

Theorem 2.4. The logarithm of the product of two numbers is equal to the sum of the logarithms of the numbers; i.e.

$$\log_a(mn) = \log_a m + \log_a n \qquad\qquad (2.14)$$

Let

$$\log_a m = x \quad \text{and} \quad \log_a n = y, \quad \text{then} \quad a^x = m \quad \text{and} \quad a^y = n.$$

Now

$$mn = a^x \cdot a^y = a^{(x+y)} \quad \text{(First Law of indices)}$$

Hence, on invoking relationships 2.12 and 2.13, we have

$$\log_a(mn) = x + y = \log_a m + \log_a n$$

Theorem 2.5. The logarithm of the quotient of two numbers is equal to the difference of the logarithms of the numbers; i.e.

$$\log_a(m/n) = \log_a m - \log_a n \qquad (2.15)$$

The proof follows the reasoning in Theorem 2.4, but the Second Law of indices is used instead of the First.

Theorem 2.6. The logarithm of a number raised to a power is equal to the power multiplied by the logarithm of the number; i.e.

$$\log_a(m^n) = n \cdot \log_a m \qquad (2.16)$$

Let

$$\log_a m = x \quad \text{then} \quad m = a^x$$

So

$$m^n = (a^x)^n = a^{xn} \quad \text{(Third Law of indices)}$$

Hence, on invoking relationships 2.12 and 2.13, we have

$$\log_a(m^n) = xn = n \cdot \log_a m$$

The above three theorems define the theory that enables logarithms to simplify arithmetical computation, by converting multiplication into an addition procedure, division into subtraction, and raising a number to a power into a multiplication process. As regards the last statement, which is associated with Theorem 2.6, it is evident that $n \cdot \log_a m$ is much more tractable than m^n. Indeed, unless n is a rational number, m^n can only be defined and evaluated (approximately) *via* the form $n \cdot \log_a m$ (see page 22).

Common logarithms

For the purpose of arithmetical calculations, logarithms to the base 10 are used. These are called Common Logarithms. The tabulation below shows why these logarithms are the most convenient in calculations.

$$\text{Since } 10^0 = 1 \qquad \text{then } \log_{10} 1 = 0$$
$$10^1 = 10 \qquad \log_{10} 10 = 1$$
$$10^2 = 100 \qquad \log_{10} 100 = 2$$
$$10^3 = 1000 \qquad \log_{10} 1000 = 3 \text{ etc.}$$

Hence all that is required is a table of logarithms to the base 10 of numbers between 1.0 and 9.9999, in which the corresponding logarithmic values lie between 0 and almost 1. The logarithm of any number $\geqslant 10$ will be given by using the table and inserting the appropriate integer in front of the decimal point, as shown in Example 2.3 below. The integer part of a logarithm is known as the characteristic, and the decimal part as the mantissa.

EXAMPLE 2.3

Find the common logarithms of

(a) 6.372   (b) 63.72   (c) 637.2

(a) The log of 6.372 can be looked up directly in the table.

$$\log_{10} 6.372 = 0.8042$$

(b) First write the number in a different form, i.e.

$$63.72 = 6.372 \times 10$$

Hence

$$\log_{10} 63.72 = \log_{10} 6.372 + \log_{10} 10 \qquad \text{(Theorem 2.4)}$$
$$= 0.8042 + 1.0000$$

Therefore

$$\log_{10} 63.72 = 1.8042$$

(c) Proceeding in the same way as in (b):

$$637.2 = 6.372 \times 100$$

Hence

$$\log_{10} 637.2 = \log_{10} 6.372 + \log_{10} 100$$
$$= 0.8042 + 2.0000$$

Therefore

$$\log_{10} 637.2 = 2.8042$$

This scheme of working can be extended to numbers between 0 and 1.

$$\text{Since } 10^{-1} = \frac{1}{10} = 0.1 \qquad \text{then } \log_{10} 0.1 \quad = -1$$

$$10^{-2} = \frac{1}{100} = 0.01 \qquad \log_{10} 0.01 \ = -2$$

$$10^{-3} = \frac{1}{1000} = 0.001 \qquad \log_{10} 0.001 = -3 \text{ etc.}$$

EXAMPLE 2.4

Find the common logarithms of

(a) 0.6372   (b) 0.06372

(a) First write the number in the form

$$0.6372 = 6.372 \times 0.1$$

Hence

$$\log_{10} 0.6372 = \log_{10} 6.372 + \log_{10} 0.1$$
$$= 0.8042 + (-1)$$

i.e.

$$\log_{10} 0.6372 = 0.8042 - 1 \qquad (2.17)$$

There are two ways of expressing this result. The common logarithm of 0.6372 can be regarded as being composed of a negative integer (−1) and a positive mantissa (0.8042). This can be expressed as:

$$\log_{10} 0.6372 = \bar{1}.8042 \qquad (2.18)$$

The minus sign is placed over the characteristic, and this implies that only the characteristic is negative and that the mantissa is positive. In words, $\bar{1}.8042$ reads: bar-1 point eight nought four two. This is the most usual way of expressing the common logarithm of a number less than 1, as it simplifies the tables necessary for converting the log of a number back to the number itself.

Returning to relation 2.17 above, we see that the negative part of the number is numerically larger than the positive part. This shows that the complete number will be negative and is, in fact, $-1.0000 + 0.8042 = -0.1958$. Hence we can also write:

$$\log_{10} 0.6372 = -0.1958 \qquad (2.19)$$

Statements 2.18 and 2.19 are both equally correct, but whereas the number on the right-hand side of the '=' sign in 2.18 is a mixed positive and negative number, that in 2.19 is a pure negative number.

(b) Proceeding as in (a), we have

$$0.06372 = 6.372 \times 0.01$$

Hence

$$\log_{10} 0.06372 = \log_{10} 6.372 + \log_{10} 0.01$$
$$= 0.8042 - 2$$

Thus we may write: either

$$\log_{10} 0.06372 = \bar{2}.8042$$

or

$$\log_{10} 0.06372 = -1.1958$$

Antilogarithms

If we know the logarithm to the base 10 of a number and we wish to know the value of the number itself, tables are available, called anti-logarithms. A table of antilogarithms is used in a similar way to a table of logarithms. The basic part of the table is for use for logarithms in the range 0.0000 to 9.9999, corresponding to antilogarithms (ordinary num-bers) in the range 1 to almost 10. The basic position of the decimal point in the antilogarithm is one place in from the left hand side of the anti-logarithm. If the logarithm contains a positive characteristic, of value n (say), then the decimal point in the antilogarithm is moved n places to the right from its basic position of one place in from this basic position. Conversely, if the logarithm contains a negative characteristic of value n, then the decimal point of the antilogarithm is moved n places to the left, and $(n - 1)$ zeros are placed between it and the first non-zero figure. Example 2.5 illustrates the method of finding antilogarithms.

EXAMPLE 2.5

Find the numbers whose common logarithms are

(a) 0.1543   (b) 2.1543   (c) $\bar{3}.1543$

(a) In the table of antilogarithms, we find that the figure in the body of the table corresponding to the margin entry of 0.1543 is 1427. Thus the antilog of 0.1543 is 1.427 (i.e. $\log_{10} 1.427 = 0.1543$).

(b) The digits of the antilogarithm are the same, 1427, but because the characteristic of the logarithm is now 2, the required antilog is 142.7 (i.e. $\log_{10} 142.7 = 2.1543$).

(c) Again, the digits of the antilogarithm are 1427, but with a characteristic of -3 ($\bar{3}$) the required antilogarithm is 0.001 427 (i.e. $\log_{10} 0.001427 = \bar{3}.1543$).

Natural logarithms

Logarithms to the base 10 are very convenient for arithmetical work. Although logarithms to any other base *could* be employed in exactly the same way, they would not be as simple to use. (This, of course, is because

10 is the base of our number system.) It was stated earlier that logarithms had other more fundamentally mathematical uses, but, in spite of being simple to use numerically, common logarithms (to base 10) do not have the simplest mathematical properties.

Logarithms which have simple and convenient mathematical properties are called Natural Logarithms, and their base is a number universally symbolized as e. The number e is irrational and its value is 2.71828, correct to 5 decimal places. It may well be asked why such a peculiar number should give rise to logarithms which are called 'natural' and have simple mathematical properties. The reasons for this will become apparent later in the book (Chapter 10), and the number e will be derived from first principles and its properties described (page 181). At present we shall only be concerned with the use of a table of natural logarithms.

A table of natural logarithms (see pages 306–7) is used in a similar way to a table of common logarithms, but there are a few differences of detail due to the fact that the base is now about 2.718 instead of 10. The characteristic of natural logs changes at integral powers of e (i.e. at $e \simeq 2.718$, $e^2 \simeq 7.389$, $e^3 \simeq 20.085$, etc.) instead of at integral powers of 10(10, 100, 1000 etc.) as in common logs. To help in determining the characteristic of a natural logarithm, this characteristic is printed in the first column of figures in the body of the table. Thus, for example, $\log_e 5.40 = 1.6864$, but we must remember that $\log_e 5.42$ (say) $= 1.6901$ and not 0.6901. Special care is needed when working in the table at places where the characteristic changes; $\log_e 2.71 = 0.9969$, whereas $\log_e 2.72 = 1.0006$ and $\log_e 2.73 = 1.0043$. Also, $\log_e 7.38 = 1.9988$, whereas $\log_e 7.39 = 2.0001$. The method of finding natural logarithms outside the range 1 to 10 is the same in principle as for common logarithms, but is not quite as straightforward in practice, as Example 2.6 shows.

It should be noted that $\log_e$ is often abbreviated to ln (remember this as '*log n*atural'). Thus

$$\log_e m = \ln m \tag{2.20}$$

EXAMPLE 2.6

Find the natural logarithms of

(a) 6.372   (b) 63.72   (c) 637.2   (d) 0.6372   (e) 0.06372

(a) This can be evaluated directly from the table.

$$\log_e 6.372 = 1.8519$$

(b) First write the number in the form

$$63.72 = 6.372 \times 10$$

Hence

$$\log_e 63.72 = \log_e 6.372 + \log_e 10$$

Now $\log_e 10$ is given in the subsidiary table at the bottom of the page (i.e. as $\log_e 10^1$), and is 2.3026. Thus

$$\log_e 63.72 = 1.8519 + 2.3026$$
$$= 4.1545$$

(c) $\qquad\qquad 637.2 = 6.372 \times 10^2$

Hence

$$\log_e 637.2 = \log_e 6.372 + \log_e 10^2$$
$$= 1.8519 + 4.6052$$

(since $\log_e 10^2 = 4.6052$ from the subsidiary table). Thus

$$\log_e 637.2 = 6.4571$$

(d) $\qquad\qquad 0.6372 = 6.372 \times 10^{-1} = 6.372 \times \dfrac{1}{10}$

$$= \frac{6.372}{10}$$

Hence

$$\log_e 0.6372 = \log_e 6.372 - \log_e 10$$
$$= 1.8519 - 2.3026$$

Thus

$$\log_e 0.6372 = -0.4507$$

This result is a pure negative number. There is never any need to quote natural logarithms of numbers less than 1 in mixed form.

(e) $\qquad\qquad 0.06372 = 6.372 \times 10^{-2} = 6.372 \times \dfrac{1}{100}$

$$= \frac{6.372}{100}$$

Hence

$$\log_e 0.06372 = \log_e 6.372 - \log_e 100$$
$$= \log_e 6.372 - \log_e 10^2$$
$$= 1.8519 - 4.6052$$

Thus

$$\log_e 0.06372 = -2.7533$$

Changing the base of a logarithm

Normally, logarithms are only tabulated to the bases of 10 and e. Occasionally, however, we require the log of a number to some other base; for example, logarithms to the base of 2 are used in formulae for species diversity in ecological work. Such logarithms can be evaluated from common logs or natural logs by using a theorem called the 'change of base theorem'.

Theorem 2.7. The logarithm of a number to a base, a, is equal to the logarithm of the same number to another base, b, divided by the logarithm of a to the base b. That is

$$\log_a m = \frac{\log_b m}{\log_b a} \tag{2.21}$$

Let

$$\log_a m = x$$

then

$$a^x = m \tag{2.22}$$

Take logarithms to base b of both sides of 2.22

$$\log_b(a^x) = \log_b m$$

i.e.

$$x \cdot \log_b a = \log_b m \qquad \text{(Theorem 2.6)}$$

and so

$$x = \frac{\log_b m}{\log_b a}$$

But

$$x = \log_a m, \quad \text{hence} \quad \log_a m = (\log_b m)/(\log_b a)$$

EXAMPLE 2.7

Evaluate $\log_2 7$

From tables we find that $\log_{10} 7 = 0.8451$ and that $\log_{10} 2 = 0.3010$. Hence

$$\log_2 7 = \frac{\log_{10} 7}{\log_{10} 2} = \frac{0.8451}{0.3010} = 2.873$$

Example 2.13 shows that the logarithm to base 2 of any number can be obtained by dividing the common log of that number by 0.3010, or alternatively by multiplying by the reciprocal of 0.3010 (= 3.322). Therefore

$$\log_2 m = 3.322 \times \log_{10} m$$

Similar 'conversion factors' may be evaluated to obtain logs to any base from logs to base 10.

The logarithm as the definition of a number raised to any power

A number such as 3^2 is perfectly comprehensible: it is 3 multiplied by itself, $3^2 = 3 \times 3 = 9$. Similarly, 3^{-2} is understandable: $3^{-2} = 1/3^2 = 1/9 = 0.\dot{1}$. Further, $3^{3/2}$ is interpreted as $\sqrt{(3)^3} = \sqrt{27} \simeq 5.19\underline{6}$. But what about such a number as $3^{2.1}$, or even 3^π?

EXAMPLE 2.8
Evaluate   (a) $3^{2.1}$ (b) 3^π
(a) This number can be expanded:

$$3^{2.1} = 3^{(2+1/10)} = 3^2 \times 3^{1/10} = 3^2 \cdot \sqrt[10]{3} = 9 \cdot \sqrt[10]{3}$$

No tables are available to evaluate tenth-root, so this part of the expression would have to be evaluated by logarithms. Splitting 2.1 up has not therefore made our task any easier: it is much the simplest plan to keep the power in its original form, and use logarithms throughout. Put

$$x = 3^{2.1}.$$

Take logarithms (to any base) of both sides:

$$\log_a x = 2.1 (\log_a 3) \qquad\qquad \text{(Theorem 2.6)}$$

Common logs are the most convenient, so

$$\log_{10} x = 2.1 (\log_{10} 3)$$
$$= 2.1 \times 0.4771$$

i.e.

$$\log_{10} x = 1.0019$$

To find x itself (= $3^{2.1}$) we need the antilog of 1.0019, which is 10.04. Therefore

$$3^{2.1} \simeq 10.04.$$

(b) Put

$$x = 3^{\pi}.$$

Take common logs of both sides:

$$\log_{10} x = \pi \cdot \log_{10} 3$$
$$= 0.4771\pi.$$

To evaluate this further we must give π an approximate value, which can be 3.142 correct to 3 decimal places. Hence

$$\log_{10} x \simeq 1.499 \quad \text{and so} \quad x \simeq 31.55.$$

Therefore

$$3^{\pi} \simeq 31.55.$$

Any number of the form a^n can be evaluated by the procedure illustrated in Example 2.8, and if n is irrational this is the *only* way in which a^n can be defined and evaluated. If n is rational, a^n can be thought of more simply in terms of powers and roots, but one may still have to use logarithms to evaluate it. It should also be noted that taking the logarithm of a number usually gives only an approximate result. Hence, both answers of Example 2.8 are approximations.

EXERCISES 2

1. Evaluate the factorials of all the integers from 1 to 10.
2. Find the values of   (a) $10!/8!$ (b) $12!/(9!3!)$.
3. What are the natural logarithms of

   (a) 7394.0   (b) 0.0438   (c) 0.00005?

4. Evaluate $0.001529^{2/9}$.
5. Show that $(\log_a b)(\log_b a) = 1$.
6. Prove that   (a) $\log_a a = 1$ (b) $\log_a a^x = x$.

3

Multi-dimensional Space

At first sight, the idea of multi-dimensional space sounds strange and forbidding, more connected with modern physics and cosmology than with biology. In fact, however, the concept of multi-dimensional space is not difficult to comprehend, and considerable biological use is made of the ideas involved, especially in the fields of ecology, genecology, and taxonomy.

LOCATION IN SPACE

Cartesian co-ordinates in two dimensions

Let us begin by considering the familiar two-dimensional space which can be represented on the plane of the paper, and consider first how to specify the position of a point on the plane. Any specification of the position of a point can only be given *relative* to the position of something else. It is most convenient to make this 'something else' a reference body of some sort, and then measure the distance of the point we require relative to this reference body. Since we are at present working in two dimensions, we require one part of the reference body to measure distances in one direction (say east-west), and another part of the reference body to measure distances at right-angles to the previous direction (i.e. north-south).

The above specification is achieved by drawing two lines at right angles to one another, as shown in Fig. 3.1. These lines are known as the **co-ordinate axes,** and the horizontal line is usually called the **x-axis** while

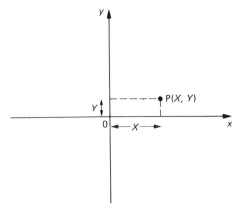

Fig. 3.1   The Cartesian co-ordinates of a point in two-dimensional space.

the vertical line is known as the **y-axis**. The position of any point on the plane may then be specified by two quantities — a distance measured along the x-axis, often called the **abscissa**, and a distance measured along the y-axis, often known as the **ordinate**. In Fig. 3.1 the point P is specified by two numbers: X, representing a definite distance along the x-axis (abscissa), and Y, which represents a definite length along the y-axis (ordinate). These two numbers, together with the name or symbol of the point in question, are specified thus: P(X, Y) and are called the **Cartesian co-ordinates** of point P, after Descartes who first specified the position of a point in this way.

The co-ordinates of the point of intersection of the axes are (0,0); this point is known as the **origin**, and is designated by 0. You will doubtless recognize the co-ordinate axes as the same as those used in elementary graphical work, but are probably more used to seeing the origin in the bottom left-hand corner of the graph instead of in the middle, as Fig. 3.1. However, the convention of measuring distances from the origin is the same in both cases: increasing distances to the right of the origin are associated with increasing positive values of x, and increasing distances above the origin are associated with increasing positive values of y. What about points to the left of and/or below the origin? Working away from the origin along the x-axis on the left-hand side we encounter increasing *negative* values of x, and as we move down from the origin along the y-axis we find increasing negative values of y.

The statements in the previous paragraph are illustrated in Fig. 3.2. In these diagrams, the intersection of the x- and y-axes at right angles delimit four quadrants, which can be named: 'upper-right', 'upper left',

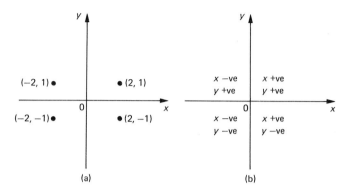

Fig. 3.2   Cartesian co-ordinates in the four quadrants defined by co-ordinate axes in two-dimensional space: (a) particular, (b) general.

'lower-left', and 'lower-right'. Figure 3.2(a) shows four points, each one being situated 2 units along the x-axis and 1 unit along the y-axis, in various directions. It can be seen that in the upper-right quadrant defined by the co-ordinate axes, values of both x and y are positive. In the upper-left quadrant, values of x are negative while those of y are positive. In the lower-left quadrant, both x and y are negative; while in the lower-right quadrant, x is positive and y is negative. These facts are summarized in Fig. 3.2(b).

As biologists, we are mainly interested in quantities which can be depicted in the upper-right quadrant — positive values of x and y. Nevertheless, we must remember that the other three quadrants exist, because, when analytical methods are applied to real data, negative values do sometimes arise.

Three dimensions

All the above ideas extend quite easily to three dimensions, and although the concept of three dimensions is everyday experience, it is difficult to illustrate three-dimensional space on paper. In attempting to do this, we use the laws of perspective and the adaptiveness of the human eye to perceive what it is asked to.

The three dimensions are conveniently termed breadth, depth, and height. If we imagine breadth and depth to be a horizontal plane (x- and y-axes), then the third dimension, height, can be represented by a vertical line (z-axis). This is illustrated in Fig. 3.3, and for simplicity we shall

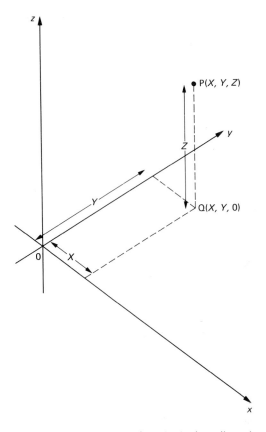

Fig. 3.3   The Cartesian co-ordinates of a point in three-dimensional space.

consider only that part of the co-ordinate system in which all quantities are positive. The situation illustrated in Fig. 3.3 can be realized in practice as a bar placed vertically at the corner of a table to represent the z-axis, and then the two edges of the table, beginning from the corner on which the bar is standing, will represent the x- and y-axes. The position of any point P on or above the table can then be specified precisely with respect to these three co-ordinate axes, as shown in Fig. 3.3. From P a perpendicular is dropped on to the $x-y$ plane; the length of this perpendicular line is Z. The point at which the perpendicular intersects the $x-y$ plane has co-ordinates (X, Y). Hence the co-ordinates of point P are (X, Y, Z).

Higher dimensions: the mathematical and physical concepts of dimension

Although a pictorial (i.e. physical) representation of the system of co-ordinates in four or more dimensions is not possible, the *mathematical* idea of a number of dimensions greater than three is perfectly feasible and sound. The concept is entirely a mathematical one and has no physical reality; the fact that we are able to realise 1, 2 and 3 dimensions in physical space is mere coincidence. The mathematical concept of dimension is distinct from the physical concept in which there is an upper limit on the number of possible dimensions. According to present knowledge the physical universe is composed of four dimensions, three of which are space, and one is time. In the mathematical sense, the number of possible dimensions is infinite.

We have been dealing with the location of points in two- and three-dimensional space, and because these dimensions in the mathematical sense can be made to accord with physical dimensions, it has been possible to give physical meanings to a set of co-ordinates for a point P, e.g. (X, Y, Z). In two dimensions the location of a point is precisely specified by two co-ordinates; similarly, in three dimensions a point is completely specified by quoting three co-ordinates. Is there any reason why we cannot add more co-ordinates to specify the position of a point in a 'space' of more than three dimensions? No reason at all. There is in fact nothing to stop us from writing the co-ordinates of a point P as say, (X, Y, Z, T, U), and we should then infer that point P was situated in a 'space' of five dimensions.

It will be convenient at this stage to change the labelling of the co-ordinate axes from that used previously. Let us now designate the x-axis as the x_1-axis, the y-axis as the x_2-axis, and the z-axis as the x_3-axis. Then, for example, the co-ordinate axes for five-dimensional 'space' will be designated $x_1, \ldots, x_5$, and the co-ordinates of point P in such a space will then be $(X_1, X_2, X_3, X_4, X_5)$.

In the last two paragraphs the word 'space' has been placed in quotation marks because we are now dealing solely with the mathematical idea of space, which is not equivalent to the familiar physical concept of space. From now on, the quotation marks will be dispensed with, although we shall always be referring to the mathematical concept of space.

In view of the fact that the mathematical idea of space has no parallel with physical space beyond the third dimension, you may be asking just what use can the biologist make of the mathematical concept of space, when he has to deal with real objects which exist in the physical sense. The example in the next section introduces what is probably the most fruitful application of the idea of multi-dimensional space to biological enquiry.

The niche of an organism in an ecosystem

An **ecosystem** is a system of inter-relationships between organisms, and between organisms and their environment. An ecosystem can be considered at any order of size and complexity, and examples are: a patch of soil in a garden, an area of woodland, or a whole island or country. Descriptively, the **niche** of an organism is the 'position' it occupies in relation to the physical factors operating, and to the other organisms present, in the environment. For instance, a particular species of plant may tolerate only an acid soil and be intolerant of competition by heather, to name but two external factors. Thus the species may find a niche on acid soil which does not carry a stand of heather.

The concept of a niche can be put into more definite quantitative terms. The niche of an organism may be defined as its position in multi-dimensional environmental space, and the mathematical ideas already discussed can be applied to the ecosystem. Each one of the co-ordinate axes represents an environmental factor, either physical or biotic, affecting the species concerned. Let us illustrate with an extremely simplified ecosystem consisting of three factors

(i) average temperature of the air, x_1;
(ii) average moisture content of the soil, x_2;
(iii) number of earthworms per unit volume of soil, x_3.

In this environmental space, organisms find their respective niches according to their physiological properties (Fig. 3.4). Organism A, which may perhaps be a species of moss, requires a high temperature, a low soil moisture content, and a reasonable earthworm population. The latter may indicate a soil having a reasonable base status. Thus, moss A could be expected to occur on dry calcareous soils in the south of England. Moss B, however, requires a lower air temperature, a much damper soil, and a small earthworm population perhaps indicating an acid soil. Moss B, then, may be found on siliceous soils in the west and north of Britain, where the climate is cool and wet.

Two points may be noted about this very over-simplified picture. Firstly, the niche of an organism would not be represented as a *point* in multi-dimensional space. Plants and animals have a tolerance range to environmental and biotic factors; this means that in our three-dimensional environment a niche will be represented by a *volume* rather than a point. The two-dimensional analogue of a volume is an *area*, while the equivalent in four or more dimensions is known as a *hyper-volume.*

Secondly, in practice the number of environmental factors is indefinitely large. Because of this, and also because we do not know what all the

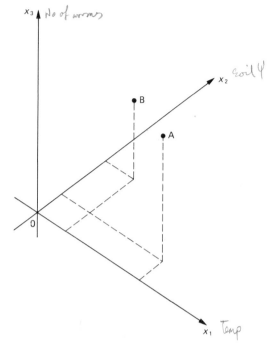

Fig. 3.4 Two points located in three-dimensional space (see page 29).

environmental and biotic factors contributing to a niche are, an ecosystem cannot be directly analysed in this way, in the present state of knowledge. The actual procedures used are much too complicated to describe here, but they are based upon the ideas and principles discussed above.

Polar co-ordinates in two dimensions

The Cartesian method of specifying the position of a point relative to co-ordinate axes is only one of several possible methods, although it is the commonest and the easiest to understand. Another system, known as **Polar co-ordinates**, is often useful in mathematics, although much more rarely in biomathematics. In this system, only one axis is required — the x-axes — on which the origin ($x = 0$) is shown, and we require to specify the position of point P in relation to the axis and the origin (Fig. 3.5). The polar co-ordinates of P are given by (r, θ), where r is the length of the

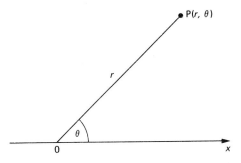

Fig. 3.5   The polar co-ordinates of a point in two-dimensional space.

straight line from the origin to P, and θ is the angle that this line makes with the x-axis. The length r is always positive, and θ can take any value between 0 and 2π radians, i.e. $0°$ and $360°$. The relations between co-ordinates in the two systems are quite simple:

Polar to Cartesian $X = r \cdot \cos \theta$ $Y = r \cdot \sin \theta$

Cartesian to Polar $r = \sqrt{(X^2 + Y^2)}$ $\theta = \tan^{-1} (Y/X)$

One use of polar co-ordinates in biological mathematics is concerned with the description of the spiral arrangement of leaves on the stems of plants — phyllotaxis.

DISTANCES BETWEEN POINTS

Two dimensions

Let $P(X, Y)$ and $Q(X', Y')$ be two points whose distance apart is required (Fig. 3.6a); that is, we want to know the length of the straight line PQ. To find out this distance, complete the right-angled triangle PQR as shown in Fig. 3.6(a). By the theorem of Pythagoras

$$PQ^2 = PR^2 + QR^2.$$

The lengths of the lines PR and QR are particularly easy to find. PR is parallel to the x-axis; its length is measured entirely in terms of x, and is $(X' - X)$. Similarly, QR is parallel to the y-axis; its length is measured in terms of y only, and is $(Y' - Y)$. Hence

$$PQ^2 = (X' - X)^2 + (Y' - Y)^2$$

and the positive square root gives the length of the line PQ.

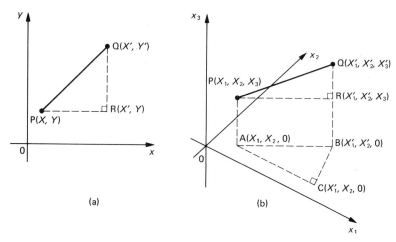

Fig. 3.6   The distance between two points: (a) in two-dimensional space, (b) in three-dimensional space.

In order to extend these ideas easily to higher dimensions, let us re-label the axes as x_1 and x_2. The co-ordinates of P will be (X_1, X_2) and of Q will be (X_1', X_2'). Then

$$PQ^2 = (X_1' - X_1)^2 + (X_2' - X_2)^2$$

or, writing $d_1 = (X_1' - X_1)$ and $d_2 = (X_2' - X_2)$

$$D^2 = d_1^2 + d_2^2 \qquad (3.1)$$

The square of the required distance, D^2, is given as the sum of two terms: the first involving a distance on the x_1-axis only, d_1, and the second term involving a distance on the x_2-axis only, d_2.

Three dimensions

Fig. 3.6(b) shows the three-dimensional case, which is analogous to the two-dimensional one we have just discussed. Two points, $P(X_1, X_2, X_3)$ and $Q(X_1', X_2', X_3')$ are located in three-dimensional space; what is their distance apart? Using exactly the same principles as before, it can be shown that

$$PQ^2 = (X_1' - X_1)^2 + (X_2' - X_2)^2 + (X_3' - X_3)^2$$

or

$$D^2 = d_1^2 + d_2^2 + d_3^2 \qquad (3.2)$$

although the proof is rather more lengthy than for the two-dimensional case. Try to work it through yourself, starting by dropping perpendicular lines from P and Q to the horizontal plane and completing the right-angled triangle ABC.

The extension from two dimensions (equation 3.1), to three dimensions (equation 3.2), is immediately apparent. We now have the distance between the two points given by three terms; the first involving a distance measured on the x_1-axis only, the second term involving a distance on the x_2-axis only, and the third term a distance on the x_3-axis only.

n dimensions

The 'D-formula' is easily extended to n-dimensions: it becomes

$$D^2 = d_1^2 + d_2^2 + d_3^2 + \ldots + d_n^2 \qquad (3.3)$$

For n dimensions the right-hand side of equation 3.3 consists of precisely n terms added together. The first term involves only distances along the x_1-axis (one dimension), the second term involves distances only along the x_2-axis (a second dimension), and so on.

Projection of a distance in n-dimensional space into a space of fewer dimensions

It is impossible to draw a graph in four or more dimensions, and coping with three dimensions graphically is not easy. It is possible, however, to gain *some idea* of the positions and distances apart of points in multidimensional space by constructing a two-dimensional graph, as we shall now show.

Referring to Fig. 3.6(a) once again, we have already remarked that the length of PR is $(X' - X)$, and that of QR is $(Y' - Y)$. Now imagine viewing the plane edgewise on, at right-angles to the x-axis; in other words, the eye is placed at the bottom of the diagram and looking upwards. From this position the eye would not discern the actual distance from P to Q; instead, the length of the line PQ would *appear* to be the length of the line PR. The length PR is called the **projection** of the length PQ on the x-axis. Similarly, the length QR is the projection of the length PQ on the y-axis. Now unless the line PQ is inclined to both the x- and y-axes at $45°$, these two projected lines will not be of the same length. However, the importance of the projection is that it enables us to represent the distance between the two points which are situated in two-dimensional space (the $x-y$ plane) on one dimension only (either the x- or the y-axis).

Similar reasoning applies to the three-dimensional case in Fig. 3.6(b).

Here we can project the distance between P and Q, which are sited in three-dimensional space, into two dimensions; e.g. on to the x_1–x_2 plane wherein the line PR or AB is the projected length, and is equal to $\sqrt{\{(X_1 - X_1)^2 + (X'_2 - X_2)^2\}}$. Or, we can go one stage further and project PQ on to a single dimension; either on to the x-axis giving a projected length AC of $(X'_1 - X_1)$, or on to the x_2-axis giving a projected length BC of $(X'_2 - X_2)$, or on to the x_3-axis giving a projected length QR of $(X'_3 - X_3)$.

The principal use of projections is to permit us to visualize, in a two-dimensional plane defined by any two of the n axes, the orientation and the length of a line joining two points situated in multi-dimensional space. Evidently, the values of the projected distances obtained differ according to which plane the projection is made on. For example, to obtain a full picture of the distance between two points in five-dimensional space, the following 10 graphs would have to be drawn: x_1 against x_2, x_1 against x_3, x_1 against x_4, x_1 against x_5, x_2 against x_3, x_2 against x_4, x_2 against x_5, x_3 against x_4, x_3 against x_5, and x_4 against x_5. Analytical methods employed, however, ensure that most of the information concerning distances between points is contained in only two or three of the above combinations of axes.

EXAMPLE 3.1

What is the distance between the following pairs of points?

(a) P(3, 5) and Q(4, 7)   (b) A(2, 3) and B(−3, 2).

(a) $PQ^2 = (4 - 3)^2 + (7 - 5)^2 = (1)^2 + (2)^2 = 5$

Hence

$$PQ = \sqrt{5} \text{ units}$$

(b) $AB^2 = \{(-3) - (2)\}^2 + \{(2) - (3)\}^2 = (-5)^2 + (-1)^2 = 26$

Hence

$$AB = \sqrt{26} \text{ units}$$

EXAMPLE 3.2

Two points, P and Q, are sited in 5-dimensional space, and their co-ordinates are (3, 5, 4, 3, 6) & (7, 8, 1, 1, 2), respectively. Find the distance between the points. Draw a graph to show the positions of the two points with respect to axes 1 and 2. What is the projected distance between the two points on the plane defined by axes 1 and 2?

$$D^2 = (3 - 7)^2 + (5 - 8)^2 + (4 - 1)^2 + (3 - 1)^2 + (6 - 2)^2$$
$$= (-4)^2 + (-3)^2 + (3)^2 + (2)^2 + (4)^2$$
$$= 16 + 9 + 9 + 4 + 16 = 54$$

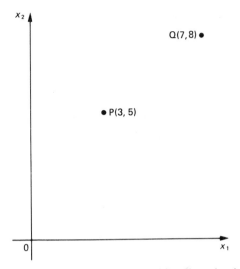

Fig. 3.7 The projection of two points situated in five-dimensional space onto the $x_1 - x_2$ axial plane (see Example 3.2).

Hence
$$D = \sqrt{54} \text{ units} \quad = 7.35$$

To plot the graph showing the points projected on to the plane defined by axes 1 and 2, we require the first two co-ordinates of each point, i.e. (3, 5) and (7, 8). The projected distance is then $\sqrt{\{(3-7)^2 + (5-8)^2\}} = \sqrt{(16+9)} = \sqrt{25}$ units (Fig. 3.7).

Quantitative characters in taxonomy

Examples of the use of distances in multi-dimensional space occur in modern taxonomy. Classical taxonomy was based as much as possible on descriptions of certain morphological features, or characters, of organisms, and the features selected were those which did not show much variation within a species or other taxonomic unit (the taxon), but did show differences between taxa. In the higher plants, floral morphology has been of particular significance. Now, however, many other morphological characters (particularly those that are measurable) are used, and special methods are available to cope with the variability that exists. Measurements such as leaf length, leaf width and internode length are made on a number of plants within a taxon, and if we consider just these three characters then

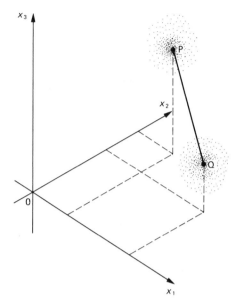

Fig. 3.8   Two clusters of points in three-dimensional space.

the measurements made can be plotted on a three-dimensional diagram (Fig. 3.8). Because of the variability from plant to plant, a cluster of points would result; but if the mean value of the measurements of each character were calculated, these three mean values would be the co-ordinates of a single point P, in the centre of the cluster. Now, take a different but related taxon and repeat the process. Another cluster of points, together with a mean point Q, would result. Evidently if the two taxa are closely related, the distance D between them would be small, and *vice versa*. The distance will be given by

$$D = \sqrt{\{(X_1' - X_1)^2 + (X_2' - X_2)^2 + (X_3' - X_3)^2\}}$$

The above example was based on the measurement of three characters per plant, thus requiring a space of three dimensions which can be represented on a diagram. There is no reason why a greater number of characters should not be measured on each plant. Each character 'requires' a dimension in space, and the distance between two taxa is calculated by the 'D-formula', using as many terms as required on the right-hand side — one term for each character measured.

If interest centres on the taxonomic relationships within a critical genus such as *Rubus*, in which the *R. fruticosus* aggregate consists of a large number of microspecies, the method of finding distances between these taxa will help to shed light on their interrelationships and will therefore be of considerable use in classifying them. In fact, we can go further. The detailed methodology, based on the above principles, for dealing with the primary data can even attempt to answer the question of whether a particular taxon should be regarded as a full microspecies or merely as a variety of another microspecies.

EXAMPLE 3.3

A certain genus of flowering plants contains 3 species, and we wish to enquire about their interrelationships. Four characters are measured on each of a number of individuals from each species, and the average for each character and species is calculated. The four characters measured, and the axis in space with which each is identified, are as follows:

x_1 — length of 3rd leaf from base of stem

x_2 — length of 3rd internode

x_3 — corolla diameter

x_4 — number of flowers per inflorescence

The results may be tabulated, thus:

	x_1	x_2	x_3	x_4
Sp. 1	2	5	2	8
Sp. 2	3	6	5	2
Sp. 3	7	4	3	7

Consider the first two characters only: the three species are shown plotted against the axes x_1 and x_2 in Fig. 3.9(a). The distances between the species are:

Sp. 1 — Sp. 2

$$D^2 = (3-2)^2 + (6-5)^2 = 1 + 1 = 2$$
$$D = \sqrt{2} \doteq 1.414$$

Sp. 2 — Sp. 3

$$D^2 = (7-3)^2 + (4-6)^2 = (4)^2 + (-2)^2 = 16 + 4 = 20$$
$$D = \sqrt{20} \doteq 4.472$$

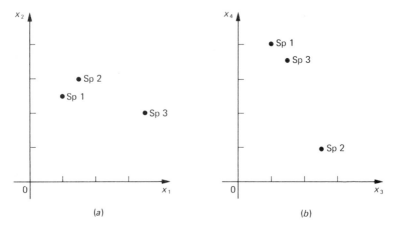

Fig. 3.9   The projection of three points situated in four-dimensional space onto:
(a) the $x_1 - x_2$ axial plane, (b) the $x_3 - x_4$ axial plane.

Sp. 3 — Sp. 1

$$D^2 = (7-2)^2 + (4-5)^2 = (5)^2 + (-1)^2 = 25 + 1 = 26$$
$$D = \sqrt{26} \simeq 5.099$$

On the basis of the two vegetative characters, from the graph and the above results, it would appear that species 1 and 2 are fairly similar while species 3 is rather different from both the other species. If now we consider the two floral characters alone, we find the situation depicted in Fig. 3.9(b). Try the distance calculations yourself, but it is clear from the graph that, on the basis of floral characters, species 1 and 3 are similar while species 2 is the outlier. Finally, let us use all the information, and calculate the inter-species distance in the four-dimensional space required to represent the information supplied. First, note the co-ordinates of the three species in four-dimensional space: Sp. 1(2, 5, 2, 8), Sp. 2(3, 6, 5, 2), Sp. 3(7, 4, 3, 7). The distances are:

Sp. 1 — Sp. 2

$$D^2 = (3-2)^2 + (6-5)^2 + (5-2)^2 + (2-8)^2 = 47$$
$$D = \sqrt{47} \simeq 6.856$$

Sp. 2 — Sp. 3

$$D^2 = (7-3)^2 + (4-6)^2 + (3-5)^2 + (7-2)^2 = 49$$
$$D = \sqrt{49} = 7.000$$

Sp. 3 − Sp. 1

$$D^2 = (7 - 2)^2 + (4 - 5)^2 + (3 - 2)^2 + (7 - 8)^2 = 28$$
$$D = \sqrt{28} \simeq 5.292$$

Now it appears that the inter-species distances are more uniform, but again, species 1 and 3 are more similar to one another than to species 2. This confirms the finding when only the two inflorescence characters were used. The plant taxonomist may wish to place more emphasis on distances involving floral characters alone, which would be in keeping with the classical approach. This, however, is for the taxonomist to decide; mathematical methods can only provide the information, but the final evaluation is firmly the province of the biologist.

LINEAR FUNCTIONS

So far, we have only considered isolated points in space, how their positions may be defined, and the distances separating them measured. These concepts, although fundamental, are limited. To extend the usefulness of mathematics we need to be able to describe and analyze 'shapes' in space.

To fix ideas, let us consider these shapes firstly in two-dimensional, and then in three-dimensional, space. In two dimensions, say on the plane of a sheet of paper, the shapes referred to above would be curved lines, or simply, **curves**. Some curves enclose an **area** − an ellipse is one such example; but most regular curves do not enclose an area within themselves.

The three-dimensional analogue of a curve is a **surface**. Some surfaces, such as an ellipsoid (e.g. the shape of a rugby football), may enclose a **volume**. For all higher dimensions the corresponding terms are **hyper-surface** and **hyper-volume**, respectively. As in the previous sections of this chapter, we shall consider the second, third, and higher dimensions separately.

Two dimensions − the straight line

In two dimensions we are working with curves. From the mathematical point of view, a straight line is a special, and indeed the simplest, type of 'curve'; and the remainder of this chapter is devoted to a study of the straight line in two-dimensional space, and of its analogues in spaces of higher dimensions.

Euclidean geometry accurately describes a straight line through two definitions: (i) a line has length but no breadth, and (ii) a straight line is the shortest line joining two points. However, the usefulness of these concepts, which define the nature of a straight line, is limited by the fact that no information on the position of the line is conveyed. If the actual location of the straight line as well as its properties can be defined, a much more complete and useful picture emerges.

The properties and position of any shape in space can be completely specified in one of two ways. One way would be to draw an accurate graph of the shape from detailed tables giving co-ordinates of the points that lie on the curve. This method of extracting information about a shape is cumbersome and time-consuming, and has severe limitations.

The second method of describing completely the properties and location of a shape does not rely on graphical presentation at all, although the latter may be a useful adjunct. The necessary information is contained in an **equation** involving the two variable quantities x and y. The general form of the equation determines the type of shape we are dealing with, and particular numerical values of the constants in the equation serve to identify a particular member of the type or class of shape. Evidently to describe a straight line, we require the **equation of the straight line**, and this can be obtained by a few algebraic manipulations based on the Euclidean properties of the straight line (referred to above), after specifying the position of the line.

Equation of the straight line which is specified by slope and intercept

There are two ways in which the position of a straight line can be defined on a plane relative to co-ordinate axes. These are: either (i) by specifying the orientation of the line with respect to the axes and the co-ordinates of any one point on it, or (ii) by specifying the co-ordinates of any two points on the line. Let us employ the first method of defining the position of a line. The orientation is given by the angle that the line makes with the x-axis, θ, and the point on the line selected for specification is that at which the line intersects the y-axis (Fig. 3.10(a)). This point is called the **intercept**, and its co-ordinates are $(0, a)$. The reason for choosing this particular point is that this specification leads directly to a form of equation known as the **standard equation of the straight line.**

The angle θ is not used directly in specifying the orientation of the line, as this would not lead to an equation involving x and y. Instead, we use the tangent of this angle and give it the symbol b. Thus $b = \tan \theta$, and this quantity is known as the **slope**, or **gradient**, of the line.

In order to derive the equation of the line shown in Fig. 3.10(a) from the specification of its position, we need to choose *any* point on the line other than the point already selected (i.e. the intercept). The co-ordinates of this other point are quite unspecific, and the point itself may be desig-

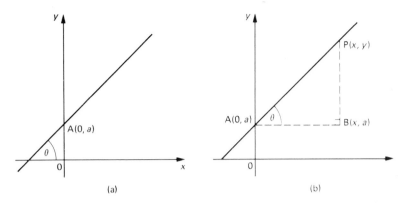

Fig. 3.10 (a) A straight line with gradient θ and intercept on the y-axis of a. (b) The same, but with construction lines added for determining the equation of the line.

nated as $P(x, y)$. This is shown in Fig. 3.10(b), together with a constructed right-angled triangle PAB. Now

$$b = \tan \theta = \frac{PB}{AB} = \frac{y - a}{x - 0}$$

Taking the purely algebraic parts from each end of the string of equalities, we have

$$b = \frac{y - a}{x}$$

Multiplying both sides by x gives us

$$bx = y - a$$

and re-arrangement finally yields

$$y = a + bx \tag{3.4}$$

Equation 3.4 is the standard equation of the straight line. The two variables, x and y, are involved; also, there are two other quantities, a and b, which specify the position of the line relative to the x–y co-ordinate axes. For any one line, a and b are **constants** with numerical values, and then the equation enables us to specify the co-ordinates of any point on the line. This is done by first specifying the x-value of the point, and then using the equation to calculate the y-value. A particular value of a and of b defines a unique line; any other pair of values defines a different line. These constants (a and b) are sometimes called **parameters**.

Equation of the straight line which is defined by slope and any point

This is a more general case of the above. We are given a straight line of gradient b which passes through a particular point $A(X, Y)$. The situation is depicted in Fig. 3.11(a): $P(x, y)$ is any point on the line other than A, and a triangle is completed as shown with point $B(x, Y)$. Now

$$b = \tan \theta = \frac{PB}{AB} = \frac{y - Y}{x - X}$$

i.e.

$$y - Y = b(x - X)$$

or

$$y = Y + b(x - X) \qquad (3.5)$$

The form of the equation 3.5 is quite useful in itself, as it shows explicitly the co-ordinates of a selected point on the line. However, the equation can easily be converted to the standard form, and this is

$$y = (Y - bX) + bx \qquad (3.6)$$

and the quantity $(Y - bX)$ is the intercept (a in our previous notation).

Equation of the straight line which is defined by any two points

It is a fundamental fact of Euclidean Geometry that any two points may be joined by a straight line. Suppose that we know the co-ordinates of two points, and are required to find the equation of the straight line joining them.

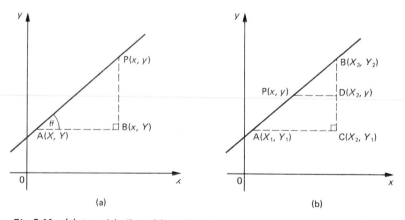

(a)   (b)

Fig. 3.11 (a) A straight line with gradient θ and passing through the point $A(X, Y)$.
(b) A straight line passing through the points $A(X_1, Y_1)$ and $B(X_2, Y_2)$.

Let the two points be A(X_1, Y_1) and B(X_2, Y_2). Draw the line joining them, and let P(x, y) be any other point on the line. Complete the construction lines as shown in Fig. 3.11(b). Now the slope of the line is given by BD/DP if triangle BDP is considered, and by BC/AC if triangle ABC is considered. Therefore, since the gradient of a straight line is the same however it is viewed, BD/DP = BC/AC. Thus

$$\frac{Y_2 - y}{X_2 - x} = \frac{Y_2 - Y_1}{X_2 - X_1}$$

After a few algebraic manipulations, this equation emerges in standard form, as follows:

$$y = \frac{X_2 Y_1 - X_1 Y_2}{X_2 - X_1} + \frac{(Y_2 - Y_1)}{(X_2 - X_1)} x \qquad (3.7)$$

Hence the slope of the line is $(Y_2 - Y_1)/(X_2 - X_1)$, which is obvious by inspection of Fig. 3.11(b), and the intercept is $(X_2 Y_1 - X_1 Y_2)/(X_2 - X_1)$.

The straight line with a negative slope
In deriving the various equations of the straight line, the diagrams have always shown the line *ascending* from left to right. The angle θ that the line makes with the x-axis, as shown in Figs. 3.10 and 3.11, lies between $0°$ and $90°$ thus yielding a positive value of tan θ. If we have a straight line which *descends* from left to right, the angle θ lies between $90°$ and $180°$ (Fig. 3.12). Now when $90° < \theta < 180°$, tan $\theta = -\tan(180° - \theta)$ (see Chapter 12, page 224); thus for lines descending from left to right, tan θ, which is b, is a negative quantity. The standard equation of the straight

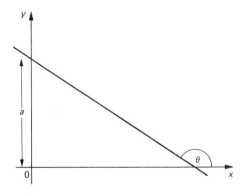

Fig. 3.12   A straight line with negative gradient.

line is still the same as equation 3.4, but b is now negative instead of positive.

The whole situation regarding the gradient of a straight line may now be summarized. When the line is horizontal, the gradient is zero. As the line moves away from the horizontal towards the vertical, sloping up to the right, the gradient increases through positive values. When the line is vertical, the gradient is infinitely large; and as the line moves away from the vertical towards the horizontal, sloping down to the right, the gradient decreases through negative values.

EXAMPLE 3.4

Find the equations of the following straight lines:

(a) slope of 1.4 and passing through the point $(0, -3.7)$;
(b) slope of -1, and passing through the point $(-2, 3)$;
(c) passing through the points $(3, 2)$ and $(-2, -3)$.

(a) Since the point $(0, -3.7)$ is on the y-axis, the intercept is -3.7. The equation can thus be written down directly:

$$y = -3.7 + 1.4x$$

(b) The slope is given as -1, and the line passes through the point $(-2, 3)$. The appropriate form of equation is given by 3.5, which in the present example becomes

$$y = 3 - (x + 2)$$

In standard form, this is

$$y = 1 - x$$

(c) First, find the gradient of the line. This is given by the difference of the ordinates of each point divided by the difference between the abscissae:

$$b = \frac{2 - (-3)}{3 - (-2)} = \frac{5}{5} = 1$$

To obtain the intercept, one can either substitute directly into equation 3.7, or the co-ordinates of one of the two points can be substituted into the standard equation. Using the first point $(3, 2)$, we have

$$2 = a + (1)(3)$$

giving $a = -1$. Hence, the equation in standard form is

$$y = -1 + x$$

Three dimensions — the plane

The equation of a straight line is a relationship between two mathematical variables, x and y. The two variables can be represented on a graph as co-ordinate axes, and the equation is depicted as a straight line whose position in relation to the axes depends on the values of the constants in the equation. The x-variable is often called *independent*, and the y-variable dependent; when we use the equation of a straight line in standard form, we are often interested in seeing what value y assumes when a particular x is selected. The idea is that y depends on x according to the equation.

When considering three dimensions, there are three variables (axes), and mathematical relationships between them in the form of equations can be

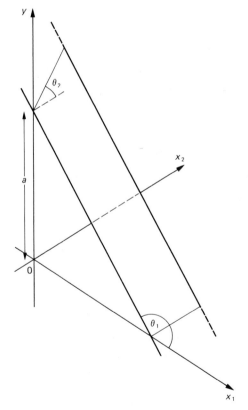

Fig. 3.13   A plane surface with gradient θ_1 on the $x_1 - y$ axial plane, gradient θ_2 on the $x_2 - y$ axial plane, and intercept on the y-axis of a.

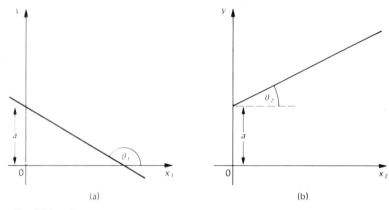

Fig. 3.14   The straight lines designating the intersections of the plane in Fig. 3.13 with: (a) the x_1-y axial plane, and (b) the x_2-y axial plane.

visualized and constructed. The 'shape' which the equation represents in three-dimensional space is a surface, which may be plane or curved. The equation itself may be considered to have two independent variables (axes), which can be designated as x_1 and x_2, and one dependent variable that can still be denoted as y.

In three dimensions, the analogue of the straight line is a plane surface, as shown in Fig. 3.13. When considering what form the equation to such a surface should take, we note that the plane intersects the y-axis; hence an intercept, a, can still be recognized. Next, consider a plane formed by the x_1 and y axes: the x_1-y axial plane. It can be seen that the plane surface crosses this axial plane as a straight line (Fig. 3.14(a)). This straight line makes an angle θ_1 with the x_1-axis, and so we may write $b_1 = \tan\theta_1$. A similar situation is found on the axial plane x_2-y (Fig. 3.14(b)): the straight line, in which the plane surface intersects the axial plane, makes an angle θ_2 with the x_2-axis, and so $b_2 = \tan\theta_2$. By a method analagous to that used for deriving the standard equation of the straight line, it can be shown that the equation of the plane surface is

$$y = a + b_1x_1 + b_2x_2 \tag{3.8}$$

You can see that this equation is very similar to the standard equation of the straight line, but because we now have two independent variables instead of one, an extra term of the form bx is added.

The straight lines shown in Fig. 3.14, which are the intersections of the plane surface with the axial planes x_1-y and x_2-y have the equations $y = a + b_1x_1$ and $y = a + b_2x_2$, respectively.

EXAMPLE 3.5

Find the equations of the following planes:

(a) gradient of 1.7 in one direction, gradient of -2.1 in the direction at right-angles to the previous direction, and passing through the point $(0, 0, 1)$;
(b) gradients of 0.5 and 1, and passing through the points $(2, -3, 1)$;
(c) having a gradient of 3 in one direction, and passing through the points $(0, 1, 1)$ and $(1, 3, 2)$.

(a) The order of the axes represented in the co-ordinate specifications is, by convention, (x_1, x_2, y). Hence the point $(0, 0, 1)$ is the intercept on the y-axis $(= 1)$, and so the equation can be written down directly:

$$y = 1 + 1.7x_1 - 2.1x_2$$

(b) Since we know the two gradients, we can first write the equation as $y = a + 0.5x_1 + x_2$. Further, we know that the point $(2, -3, 1)$ lies on the surface; so, substituting into the preliminary equation, we have

$$1 = a + (0.5)(2) + (-3)$$

i.e.

$$1 = a + 1 - 3 \quad \text{giving } a = 3$$

So the final equation is

$$y = 3 + 0.5x_1 + x_2$$

(c) Since the information given about the gradient is not specific about the direction, we can write two possible preliminary equations:

$$y = a + 3x_1 + b_2x_2$$

or

$$y = a + b_1x_1 + 3x_2$$

We are also given the co-ordinates of two points that lie on the surface; so, using one of the above preliminary equations, we can form two equations in two unknowns — the intercept, a, and one of the gradients. Using the first of the preliminary equations, we have

$$1 = a + 3(0) + b_2(1)$$

and

$$2 = a + 3(1) + b_2(3)$$

and solving this pair of equations gives $a = 2$ and $b_2 = -1$, giving the equation of the plane surface as

$$y = 2 + 3x_1 - x_2$$

Substituting the co-ordinates of the two points successively into the second preliminary equation gives

$$1 = a + b_1(0) + 3(1)$$

and

$$2 = a + b_1(1) + 3(3)$$

which yield the solution $a = -2$ and $b_1 = -5$. So now the equation of the plane surface is

$$y = -2 - 5x_1 + 3x_2.$$

If three points had been specified, and no information about any gradient given, it would have been necessary to solve three simultaneous equations in three unknowns, unless one of the given points was the intercept when we should have two equations in two unknowns.

n dimensions — the hyperplane

As usual, more than three dimensions cannot be illustrated; but, as we saw when specifying the position of a point, and calculating distances between points, formulae established for two and three dimensions extend quite easily to higher dimensions.

In the present situation we have n dimensions, which means $(n - 1)$ independent variables $(x_1, \ldots, x_{(n-1)})$ and one dependent variable (y). The analagous equation to the two-dimensional straight line (3.4) and the three-dimensional plane surface (3.8) is

$$y = a + b_1x_1 + b_2x_2 + b_3x_3 + \ldots + b_{(n-1)}x_{(n-1)} \qquad (3.9)$$

Any axial plane formed by the y-axis and any one x-axis, x_i, will have an equation of the form

$$y = a + b_ix_i \quad i = 1, \ldots, (n - 1) \qquad (3.10)$$

Expression 3.10 means that i can have any value from 1 up to $(n - 1)$ inclusive. For instance, if $i = 5$, equation (3.10) becomes $y = a + b_5x_5$ and represents the straight line given by the intersection of the hyperplane with the axial plane $x_5 - y$.

Linear functions

The most general equation that we have encountered in this section of the chapter is equation 3.9; the other similar equations, 3.4 and 3.8 are simply sub-sets of the right-hand side of 3.9. The two most important features of equation 3.9 are: (i) that all the variables, the x_i and y, are of degree one (i.e. there is no term such as y^2 or x_4^3); and (ii) that only one x_i or y appears in any term (i.e. there is no term of the form $x_i y$, or $x_i x_j$ $i \neq j$). This kind of equation is known as a **linear** equation, because the simplest equation of this type is that of a straight line.

Therefore, a linear equation in two variables is the equation of a straight line. A linear equation in three variables is the equation of a plane, while a linear equation in four or more variables is the equation of a hyperplane.

Instead of the word 'equation', the term 'function' (short for 'mathematical function') is more usual. Thus, in the equation of a straight line, $y = a + bx$, we say that y is a linear function of x. In the equation of a plane, $y = a + b_1 x_1 + b_2 x_2$, we say that y is a linear function of x_1 and x_2; and so on. In describing the relationship this way, we again emphasise the dependent nature of the variable y, making it a function of one or more x_i which are given.

There are, of course, many types of mathematical functions other than linear ones, and we shall meet some useful ones in the next and later chapters.

LINEAR FUNCTIONS IN BIOLOGY

Choosing and fitting a function

The use of a mathematical function in biology is to provide a quantitative description of the way some biological activity proceeds in relation to some factor external to the biological system. Some examples are: the rate of photosynthesis in relation to light intensity, the growth of a population with respect to time, the change in size of some character of a species of organism in relation to altitude above sea level. In all these examples, the x-axis would represent the external factor — light intensity, time, altitude; and the y-axis represents the biological **response** — photosynthetic rate, number of organisms in a population, size of a particular character in a certain species.

In examples of this kind, one is interested in a possible mathematical relationship (in the form of a function) between the two variables, to describe the biological response to changes in the external factor. This raises two questions: (i) what kind of mathematical function is involved and, when this question has been answered, (ii) what are the values of the constants, or parameters, of the function? In some instances, which are rare in biology, theoretical considerations can provide the answer to the first question. More commonly, however, one must find a function which reasonably describes the situation by carrying out a suitable experiment, plotting the observations on a graph, and examining for trends. A function selected in this way gives an **empirical**, rather than a theoretical, description of the biological process.

From the remarks made towards the end of Chapter 1 (page 5), it is obvious that the experimental observations will not lie *exactly* on the curve of any function, and so the choice of function becomes more difficult. In many situations, however, a linear function provides an adequate empirical description of the response of a biological process to variations in level of an external factor. It is considerably easier to detect by eye deviations from a straight line than deviations from any other type of curve, and so linear functions find considerable use in biology.

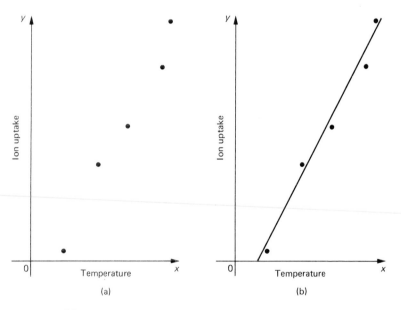

Fig. 3.15   (a) An hypothetical set of data from an experiment on the response of uptake of an ion by plant tissue at different temperatures. (b) The same, but with a 'fitted' straight line (see page 51).

As an example, consider an experiment to assess the way in which the uptake of an ion by plant tissue is affected by temperature. The experiment would consist of immersing samples of the tissue in a series of vessels holding a solution containing the ion, each vessel being maintained at a different temperature. After a certain fixed interval of time the plant tissue is fixed and assayed for the content of the ion, and the results (amount of the ion taken up at each temperature) are plotted on a graph, as in Fig. 3.15(a). Plotted in this way, the data suggest a straight line.

When the type of function has been decided upon, the constants of the function have to be estimated for the particular set of data in hand. For our present example, we have selected a linear function as suitable, and we now require values for the constants a and b of a straight line that will best describe the data shown in Fig. 3.15(a). The process of estimating the constants of a function is termed **fitting a function**, in general; and in this example, where the function to be fitted is linear, the process is usually described as **fitting a straight line** to the data.

The fitting process requires a knowledge of statistics, and so is outside the scope of this book; but let us assume that a straight line is now fitted to the data of Fig. 3.15(a), and that the line is drawn on the graph by finding the co-ordinates of two points on the line and joining them up. The final graph now appears as in Fig. 3.15(b).

The uses of a fitted function

Broadly speaking, there are three uses of a fitted mathematical function: (i) smoothing of the data, (ii) prediction purposes, and (iii) use of the mathematical function itself (i.e. its form and parameter values) to provide an insight into the working of the process under study. The only use of (i) is to provide a clearer picture of the trend involved when presenting the results graphically.

Prediction means specifying a value of x, and asking what the corresponding value of y is likely to be. In the present example, one can enquire about the most likely quantity of ion taken up at a given temperature. A word of warning is necessary, however, as one can either make predictions within the range of the data (**interpolation**) or outside the range of the data (**extrapolation**). The range of the data in the example of ion uptake extends from the lowest temperature employed *in the original experiment* to the highest temperature used. It is safe to interpolate, but not to extrapolate. The reason is that a linear function has been fitted to actual experimental data on the basis that, over the range of temperatures covered by the experiment, such a function appears to provide a reasonable description of the situation. It is certainly not safe to assume that this linear function will adequately describe the situation at lower or higher temperatures than those embraced by the experiment. Indeed, common sense tells us that at very high temperatures the whole biological system breaks down, and could certainly no longer continue in a linear mode.

If one has empirically fitted a function to a set of experimental data, then the use described in (iii) above should be employed with great caution. If one has theoretically deduced the existence of a particular function, before fitting the function to experimental data, then the constants can be used, perhaps (in the ion uptake example) to compare the uptake of different ions by the same plant species, or to compare the uptake of the same ion by different species. Even when the function is an empirical description of the data, comparisons of this kind may be made, so long as nothing more fundamental is attributed to the parameters of the equation; for in this case the parameter values found will simply be characteristic of the particular experiment that was carried out.

Extension to more than one external factor

The above type of analysis can be extended to cover more than one external factor, and then a space of more than two dimensions will be required. In the ion uptake example, we might be interested in the effects of both temperature and pH; so these two factors would be identified with axes x_1 and x_2 of a three-dimensional graph, while the uptake of ion remains as the y-axis. Assuming that a linear relationship also holds over a restricted range of pH with regard to ion uptake, the entire situation can be described by equation 3.8. In theory, there is clearly no limit to the number of external factors than can be included in a single investigation; but in practice, it may be impossible to include very many external factors in the same investigation because of the large size of the experiment needed, and the subsequent difficulty of interpreting the results.

EXERCISES 3

1. Find the distances between the pairs of points whose co-ordinates are:
   (a) $(-1, 3)$ & $(2, -4)$
   (b) $(0, 3, 4)$ & $(-2, -1, 5)$
   (c) $(1, 1, -1, 2)$ & $(-2, -3, 4, 4)$

2. The following table gives details of various morphological measurements made on common British members of the genus *Veronica* (Speedwell). All measurements are in mm, and each figure is the arithmetic mean of a number of individuals.

Species	Leaf length	Leaf width	Petiole length	Internode length	Pedicel length	Corolla diameter
V. beccabunga	35.00	17.58	3.42	68.00	29.92	6.83
V. officinalis	14.80	9.40	3.10	20.80	19.90	5.20
V. montana	22.19	17.56	9.25	38.25	38.63	8.94
V. chamaedrys	16.30	14.46	1.06	29.12	26.21	11.39
V. serpyllifolia	5.40	2.84	0.44	3.56	2.56	4.84
V. hederifolia	8.58	9.70	2.67	7.58	5.61	5.21
V. persica	11.13	10.70	1.83	6.83	17.00	9.87
V. filiformis	5.49	6.85	1.97	6.82	19.52	11.67

Draw graphs to show the positions of the species with respect to axes representing various pairs of characters. Also, calculate the distance between the species, as characterised by the above figures, in the full 6-dimensional space. Compare your answers with the taxonomic arrangement of the species in genus *Veronica* in a standard Flora.

3. Find the equations of the following straight lines:
   (a) making an angle of $60°$ with the x-axis, and passing through the point $(2, 1.3495)$;
   (b) passing through the points $(-1, 0)$ and $(5, 4)$.

4

Functions and Curves

In the previous chapter, we discussed the idea that a straight line in a two-dimensional plane could be completely described by an equation, $y = a + bx$, where x and y are the co-ordinate axes. For various reasons, such an equation is best called a mathematical function; in this instance, y is a function of x. The equation also contains two constants, a and b, and any particular pair of values of these constants defines a unique straight line. When we come to examine the *form* of the straight line function, we see that it consists of three terms: one of the terms contains the variable x, another the variable y and the third term contains neither x nor y but is merely a constant. Moreover, there is no term containing x and y together, neither are there any terms in which x and/or y are raised to a power (other than to the power 1, of course). All these properties serve to define the form of the equation of a straight line, and so we know that any equation with these properties will be that of a straight line. For example, the equation $x - 2y + 3 = 0$ represents a straight line because its form fulfils all the requirements listed above. This equation can be rearranged into a more familiar form: $y = \frac{1}{2}x + \frac{3}{2}$; which shows that the gradient of the line is $\frac{1}{2}$ and that the intercept is $1\frac{1}{2}$.

A mathematical function whose form differs in *any* way from the specification contained in the previous paragraph will not represent a straight line, but a **curve**. Any particular form of function gives rise to a unique type of curve, and the number of different functional forms and their corresponding curves is indefinitely large. Fortunately, however, many types of function and their curves can be grouped into 'families' and, as biologists, we need only concern ourselves with a few of these families.

POLYNOMIAL FUNCTIONS

The standard equation of a straight line is $y = a + bx$: on the right-hand side there is a term consisting of a constant only, and a term consisting of the variable x multiplied by a constant. A constant which multiplies a variable is called a **coefficient**. If another term of similar form, consisting of x^2 and a coefficient, is added, the above equation becomes $y = a + bx + cx^2$. Other terms could be added to the right-hand side, in which the power of x is increased by one for each new term added. Thus we have $y = a + bx + cx^2 + dx^3$ and $y = a + bx + cx^2 + dx^3 + ex^4$, for example, where a, b, c, d, and e are constants, and the number of equations that can be set up by this process is clearly infinite. Evidently, all the functions of this type can be regarded as comprising a family, and collectively they are known as **polynomial functions**, or just **polynomials**.

Nomenclature

The names applied to the first few members of the polynomial family are summarised in the table below.

Equation	Polynomial name	Special name
$= a + bx$	first degree polynomial	linear function
$= a + bx + cx^2$	second degree polynomial	quadratic function
$= a + bx + cx^2 + dx^3$	third degree polynomial	cubic function
$= a + bx + cx^2 + dx^3 + ex^4$	fourth degree polynomial	quartic function
$= a + bx + cx^2 + dx^3 + ex^4 + fx^5$	fifth degree polynomial	quintic function

Second degree polynomial

Curves of the equation $y = a + bx + cx^2$ are shown in Fig. 4.1. The actual position of the curve relative to the co-ordinate axes will, of course, depend on the values of a, b, and c. But the orientation of the curve, that is, whether the open end faces upwards or downwards, depends only

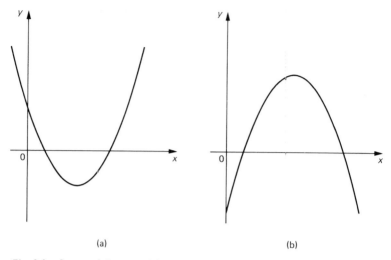

Fig. 4.1   Curves of the second degree polynomial function, $y = a + bx + cx^2$ with real roots: (a) c positive, (b) c negative.

upon the sign of c, the coefficient of x^2; in other words, upon the coefficient of the term containing the highest degree of x in the equation.

In a quadratic function wherein c is positive, working from left to right along the x-axis, the curve descends to a minimum value of y and subsequently rises again (Fig. 4.1(a)). The curve is symmetrical about a vertical line drawn through the lowest point. A curve in which c is negative rises to a maximum value and then descends (Fig. 4.1(b)); it has the same kind of symmetry as the curve with c positive.

The roots of the second degree polynomial

At the points where the curve intersects the x-axis, $y = 0$, which gives the equation

$$a + bx + cx^2 = 0 \tag{4.1}$$

the (once!) familiar quadratic equation of school-days. Solving this equation may yield two values of x, which specify the points of intersection of the curve with the x-axis. These x-values are called the **roots** of the equation 4.1, and they can be evaluated using the formula

$$x = \frac{-b \pm \sqrt{(b^2 - 4ac)}}{2c} \tag{4.2}$$

with the coefficients as defined in equation 4.1.

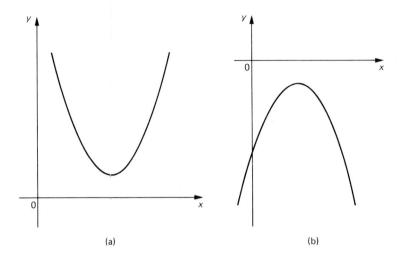

Fig. 4.2   Curves of the second degree polynomial function $y = a + bx + cx^2$ with non-real roots: (a) c positive (b) c negative.

A quadratic curve does not necessarily intersect the x-axis. In equation 4.2, the quantity within the square root sign, $(b^2 - 4ac)$, is known as the **discriminant** of equation 4.1. If $b^2 > 4ac$, then the discriminant is positive and has a real square root. Thus the two points of intersection of the curve and the x-axis are at $x = \{-b + \sqrt{(b^2 - 4ac)}\}/2c$ and at $x = \{-b - \sqrt{(b^2 - 4ac)}\}/2c$, and the situation is as shown in Fig. 4.1. If, however, $b^2 < 4ac$, the discriminant is negative and the square root is a complex number (page 9). Under these conditions the quadratic equation has no real roots, and the curve does not intersect the x-axis (Fig. 4.2). The intermediate situation is when $b^2 = 4ac$, giving a discriminant of zero. Now the curve just touches the x-axis at one point only, given by $x = -b/2c$.

EXAMPLE 4.1

Find the roots, if real, of the following equations

(a) $2x^2 - 7x + 4 = 0$
(b) $5x^2 - 4x + 2 = 0$
(c) $x^2 - 6x + 9 = 0$

(a) The left-hand side of the equation is the opposite way round from that quoted in equation 4.1, so be careful when matching the coefficients; it is the powers of x in the various terms that matter, and not the order of appearance of the terms from left to right. So $a = 4$, $b = -7$, $c = 2$. Then

$$x = \frac{7 \pm \sqrt{(49 - 32)}}{4} = \frac{7 \pm \sqrt{17}}{4} = \frac{7 \pm 4.123}{4}$$

So either

$$x = \frac{7 + 4.123}{4} = \frac{11.123}{4} = 2.78$$

or

$$x = \frac{7 - 4.123}{4} = \frac{2.877}{4} = 0.72 \quad \text{(correct to 2 decimal places)}$$

(b) Here, we have $a = 2$, $b = -4$, $c = 5$. So

$$x = \frac{4 \pm \sqrt{(16 - 40)}}{10}$$

The discriminant is negative, and so there are no real roots.

(c) In this example, $a = 9$, $b = -6$, $c = 1$. So

$$x = \frac{6 \pm \sqrt{(36 - 36)}}{2} = 3$$

and the equation has only one root or, as is sometimes said, two coincident roots.

Third degree polynomial

Curves of the function $y = a + bx + cx^2 + dx^3$ are shown in Fig. 4.3. As in the case of the quadratic curve, the actual position of a cubic curve relative to the co-ordinate axes depends on the values of the four constants: a, b, c, and d. There are two different orientations (Fig. 4.3(a), (b) and (c), (d)) according to the sign of d, the coefficient of x^3 (i.e. the coefficient of the term containing the highest power of x in the equation). Moreover, there may or may not be a loop in the curve (Fig. 4.3(a), (c) and (b), (d)), again dependent on the sizes of the coefficients.

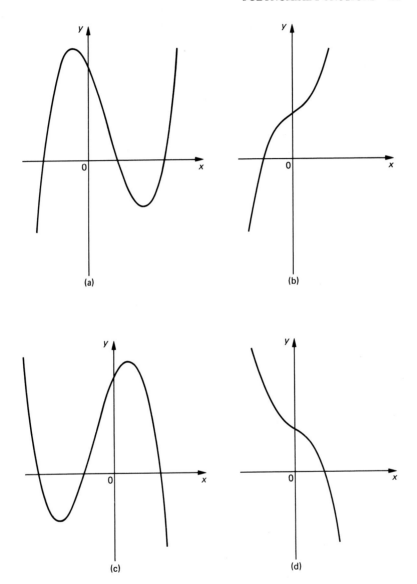

Fig. 4.3 Curves of the third degree polynomial function $y = a + bx + cx^2 + dx^3$: (a) and (b) d positive, (c) and (d) d negative.

At a first reading, the following sub-section can be left-out and you may proceed directly to the section 'Polynomials in Biology' on page 62.

The roots of the third degree polynomial

Solution of the equation

$$a + bx + cx^2 + dx^3 = 0 \qquad (4.3)$$

specifies the point or points of intersection of the curve with the x-axis. In any cubic curve there will always be one such point, and there may be three if the curve contains a loop which crosses the x-axis. The intermediate situation of two roots is that in which the bottom or top of the loop, as the case may be, just touches the x-axis.

While equation 4.3 *can* be solved explicitly, the method of doing so is not very straightforward. There are several stages in the solution, and the final step of actually obtaining the roots is too involved to be presented here. However, we shall describe how to obtain the discriminant of the equation so that the number of real roots is at least known, and they can then be approximated by the iterative method described in Chapter 6.

First, the equation has to be obtained in a so-called **standard form**; in this form there is no term in the square of the variable, and the coefficient of the cubic term is unity. Putting the equation into standard form involves defining a new variable, z, which is formed by adding $c/3d$ to the original variable, x, and so $z = x + c/3d$ or $x = z - c/3d$. Then we substitute for x in the original equation (4.2):

$$a + b\left(z - \frac{c}{3d}\right) + c\left(z - \frac{c}{3d}\right)^2 + d\left(z - \frac{c}{3d}\right)^3 = 0$$

From this equation we obtain, after some algebra:

$$\frac{1}{d^3}(ad^2 - bcd + c^3 d - c^3) + \frac{1}{d^2}(bd - c^2)z + z^3 = 0 \qquad (4.4)$$

Notice the absence of a term in z^2 and that the coefficient of z^3 is 1. Now let $p = (bd - c^2)/d^2$ and $q = (ad^2 - bcd + c^3 d - c^3)/d^3$ then equation 4.4 can be written as

$$z^3 + pz + q = 0 \qquad (4.5)$$

Equation 4.5 is the standard cubic equation, and the quantity $p^3 + 27q^2$ is the discriminant. If the discriminant is negative, equation 4.5 has three real roots, if positive there is only one real root, and if the discriminant is zero we have the intermediate condition of two real roots, i.e. two coincident roots and one other.

In obtaining the standard form of the equation, we obtained a new

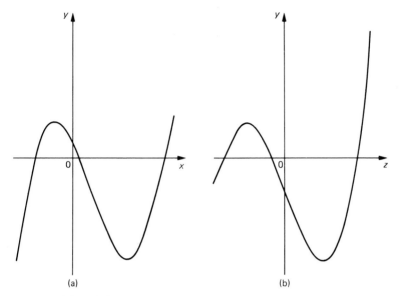

Fig. 4.4 (a) and (b) Two third degree polynomials differing only in their positions on the x-axis.

variable, z, merely by adding a constant, $c/3d$, to the original variable, x. This means that the cubic in z, equation 4.4, has a curve of exactly the same shape and bears exactly the same relationship to the y-axis as does the original equation 4.3; there is merely a displacement along the horizontal axis, and no displacement vertically. Hence, the number of roots established by the discriminant for the standard form of the equation will be the same as for the original. The situation is shown in Fig. 4.4.

EXAMPLE 4.2

How many roots has the cubic equation

$$x^3 - 2x^2 - 4x + 1 = 0? \tag{4.6}$$

Matching the coefficients with the symbolic coefficients of equation 4.3: $a = 1, b = -4, c = -2, d = 1$. Thus $c/3d = -2/3$, and so $z = x + (-2/3) = x - 2/3$ and $x = z + 2/3$. Substituting for x in (4.6) gives

$$\left(z + \frac{2}{3}\right)^3 - 2\left(z + \frac{2}{3}\right)^2 - 4\left(z + \frac{2}{3}\right) + 1 = 0$$

After multiplying out the brackets and collecting together like terms, we have

$$z^3 - \frac{16}{3} z - \frac{61}{27} = 0 \qquad (4.7)$$

Equation 4.7 is the standard form of equation 4.6. Next, compare the coefficients of equation 4.7 with those in the general standard cubic, equation 4.5: $p = -16/3$, $q = -61/27$. The discriminant is $p^3 + 27q^2 = -4096/27 + 3721/27 = -375/27$. Therefore, as the discriminant is negative, equation 4.7, and so equation 4.6, has three real roots.

Polynomials in biology

Polynomial functions and their curves are very important in mathematics, as they have relatively simple mathematical properties. Accordingly, we shall make considerable use of polynomials in succeeding chapters to illustrate ideas and methods. On the other hand, in the mathematical description of biological phenomena polynomials find little place; but they may be useful for smoothing and forecasting purposes.

THE FUNCTION ac^x

Polynomial functions comprise a series of terms in which x is raised to one or more integer powers. In this section we shall examine a family of functions in which x itself is the power, i.e.

$$y = ac^x \qquad (4.8)$$

where a and c are constants.

In order to be able to draw a graph of the function, we need to evaluate y for a number of selected values of x. This involves using logarithms in the way described at the end of Chapter 2, unless integer values of x only are used, which is inconveniently restrictive. Taking logarithms (to any convenient base) of both sides of equation 4.8, we obtain

$$\log y = \log(ac^x)$$

Using the result of Theorem 2.4 on page 14, the right-hand side can be recast:

$$\log y = \log a + \log(c^x)$$

and finally, applying the result of Theorem 2.6, we have

$$\log y = \log a + x \cdot \log c \qquad (4.9)$$

Equation 4.9, which has been derived from the original function 4.8 merely by taking logarithms of both sides, is particularly interesting. Firstly, notice that the variable x no longer occurs as a power, neither is it raised to any power. Secondly, the variable y has been replaced by $\log y$, which appears on its own on the left-hand side. Thirdly, observe that x is multiplied by a constant, $\log c$ (remember that c is a constant), and that the remaining term is also a constant, $\log a$. Evidently, equation 4.9 represents a straight line; no longer a linear relationship between the variables y and x, but instead between the variables $\log y$ and x. This means that if we determine the values of $\log y$ for each of a number of x-values, using equation 4.9, and if we then plot $\log y$ against x on a graph, we shall obtain a straight line (Fig. 4.5(c) & (d)). The gradient of the line is, of course, $\log c$, and the intercept is $\log a$. If $\log c$ is positive, the line slopes up to the right, and *vice versa*.

To obtain the graph of the original function, $y = ac^x$, we need only add one extra step to those outlined in the previous paragraph, namely to take the antilogarithm of each value of $\log y$. This gives corresponding y-values for the selected x-values, enabling the graph of y against x to be plotted. The latter graph takes two forms: if $c > 1$, the curve slopes steadily up to the right (Fig. 4.5(a)); while if $c < 1$, the curves descends steadily from the left (Fig. 4.5(b)). The intercept in each case is a, as may be verified by putting $x = 0$ in equation 4.8: $y = ac^0$, i.e. $y = a$, since $c^0 = 1$.

Other properties of the curve of the function $y = ac^x$ are best obtained by examining the properties of the straight-line form of the relation (equation 4.9), and converting $\log y$ into y by taking antilogarithms.

The most fundamental aspect of the function $y = ac^x$ is that both the constants, a and c, are positive numbers; otherwise, in equation 4.9 we would have the logarithms of negative numbers, which are not defined (page 14). The same reasoning implies that y must always be a positive quantity too. So although the curve approaches the x-axis (i.e. y becomes small) when x takes large negative values if $c > 1$, or when x takes large positive values, if $c < 1$, the curve never actually intersects the x-axis, nor even touches it. This kind of behaviour is an extremely important mathematical phenomenon, and it is very relevant in the mathematical description of biological processes: it is a topic that will recur in greater detail later in this chapter, and at intervals throughout the book.

In a function where $c < 1$, $\log c$ will be negative, which explains why the straight line depicting the function $\log y = \log a + x \cdot \log c$ slopes down to the right. In such a function, increasing values of x imply decreasing values of $\log y$, and hence of y. This explains why the curve of the function $y = ac^x$, where $c < 1$, descends to the right. The converse reasoning applies to these properties where $c > 1$. If $c = 1$, we have $y = a \cdot 1^x$. Now 1 raised to any power is always 1, thus $y = a$. Again, $\log y = \log a + x \cdot \log 1$;

log 1 (to any base) equals 0, so log y = log a. In other words, the graph of either equation 4.8 or 4.9 when $c = 1$ is a horizontal line at height a or log a, respectively, above the x-axis.

Finally, although the intercept a of the function $y = ac^x$ must be positive, if its value is less than 1 the intercept of the function

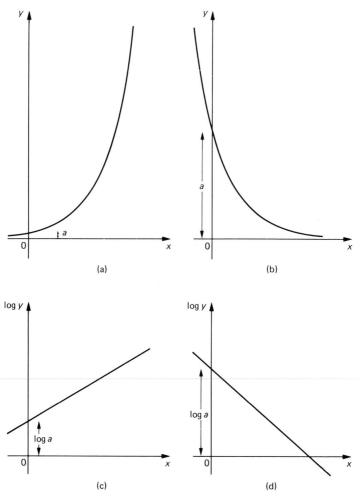

(a)

(b)

(c)

(d)

Fig. 4.5   The function $y = ac^x$: (a) curve of the function when $c > 1$, (b) curve of the function when $c < 1$, (c) the straight line log y = log $a + x \cdot$ log c corresponding to (a), (d) the straight line corresponding to (b).

$\log y = \log a + x \cdot \log c$ will be negative, since the logarithm of a number less than 1 is negative.

The *general* family of functions represented by equation 4.8 is not of particular relevance in biological mathematics, but there is one member of the family which is of outstanding importance. This particular function is obtained when $c = e$, the base of natural logarithms, and is known as the **exponential function**. For this reason, and because the function $y = ac^x$ has many properties which are of fundamental importance to the biological mathematician, the function has been extensively discussed above. The exponential function will be fully described in Chapter 10.

FUNCTIONAL NOTATION

In this chapter, two types of mathematical function have been examined. Let us take as an example for the present discussion the second degree polynomial:

$$y = a + bx + cx^2 \tag{4.10}$$

Now the essence of this function lies on the right-hand side of the equals sign, since the quadratic function represented in 4.10 is a function *of x*. In fact, we only invoke the other variable, y, for convenience so that we may draw a graph of the function on the usual $x-y$ axes. A way of writing equation 4.10 without the use of y is

$$f(x) = a + bx + cx^2 \tag{4.11}$$

and the left-hand side of 4.11 is read: 'function of x', or simply 'function-x'.

One important feature of a given mathematical function of x is that it can normally be evaluated numerically for any given value of x. Let us consider the function

$$f(x) = 3 + 2x - x^2$$

If

$$x = 0 \qquad f(0) = 3 + 2(0) - (0)^2 = 3$$
$$x = 1 \qquad f(1) = 3 + 2(1) - (1)^2 = 4$$
$$x = 5 \qquad f(5) = 3 + 2(5) - (5)^2 = -12$$
$$x = -2 \qquad f(-2) = 3 + 2(-2) - (-2)^2 = -5$$

There is another, equally important, use of the functional notation. In many situations, a mathematical function is known or inferred to exist between two variables, but the actual form of function may be in doubt or completely unknown. Under these circumstances, we may write

$$y = f(x) \tag{4.12}$$

which implies the existence of a mathematical functional relationship between x and y, but does not specify its type. The letter f in the notation could be some other letter; thus $y = f(x), y = F(x), y = \phi(x)$ denote three different functional relationships between x and y. The notation can also be extended to cover the situation where y is a function of more than one variable. For example, we may write

$$y = f(x_1, x_2) \tag{4.13}$$

which implies that y is a function of both x_1 and x_2 (c.f. page 46).

SOME PROPERTIES OF CURVES

The gradient of a curve

When discussing the straight line, we found that its orientation (slope or gradient) contributed towards its definition. The gradient of a straight line is the tangent of the angle the line makes with the horizontal axis, measured in an anti-clockwise direction; this quantity is given the symbol b and it appears as the coefficient of x in the equation describing the line.

A straight line is, however, a single special case of the more general phenomenon of curves. Can the idea of a gradient, which is an intuitively obvious property of a straight line, be extended so that it may be applied to a curve? In other words, can we speak of 'the gradient of a curve', and if so, how is it defined? The concept of the gradient of a curve is indeed valid, and is approached through two definitions.

(a) The **tangent to a curve** *at a particular point* is defined as that straight line which just touches, but does not intersect, the curve at the point under consideration.

(b) The **gradient of a curve** *at a particular point* is defined as the gradient of the tangent of the curve at that point.

In Fig. 4.6(a) is shown a curve of some function $y = f(x)$ with tangents at two points, P and Q. It is impossible in a diagram to show that a straight line can touch a curve at one definite point only, but this is only because a drawn curve and line have a finite thickness. The strict geometrical definitions of lines and curves state that they are infinitely thin. However, it is obvious that the tangent to the curve at point P is not the same as the tangent to the curve at point Q; the distinction is that the *gradients* of the two tangents differ. The gradient of a straight line is the same wherever it

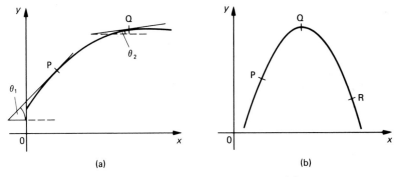

(a)   (b)

Fig. 4.6   The gradient of a curve: (a) showing the tangents, (b) without tangents.

is examined, but the gradient of a curve varies from point to point along
the curve.

Although the gradient of a curve has been defined in terms of the
gradient of a tangent to the curve, we can always think of the gradient of
a curve without a direct reference to a tangent. In Fig. 4.6(b), it is evident
that the gradient of the curve at point P is positive because the tangent at
that point has a positive gradient; the gradient at point R is negative for a
similar reason, and the gradient of the curve at point Q is zero because the
tangent at that point is a horizontal line (gradient zero).

Hence, the concept of the gradient of a curve is a valid one, and for a
particular curve whose equation is known it is possible to derive a formula
which enables us to evaluate the gradient of the curve at any desired value
of x. The gradient of a curve and its evaluation form the basis of the differential calculus.

Asymptotes

An **asymptote** is formally defined as a tangent to a curve at infinity. In
more diagrammatic, but less accurate, terms it is a line towards which a
curve may tend, but which the curve never intersects or meets. We have
already encountered such a situation in the curve of the function $y = ac^x$.
Consider the case where the constant c lies in the range $0 < c < 1$, when
the curve of the equation will be as shown in Fig. 4.5(b). With increasing
values of x, y decreases but never becomes zero or negative. Hence on the
graph as one moves further over to the right-hand side, the curve descends
towards the x-axis but never actually touches or intersects it. Thus, the
x-axis is an asymptote to the curve of $y = ac^x$ when $0 < c < 1$. Similarly,

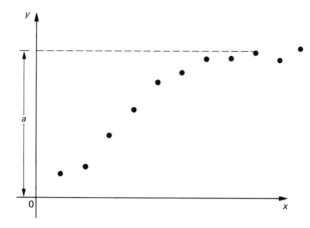

Fig. 4.7   Data requiring the fitting of a function with an asymptote $y = a$.

the curve of this function when the constant c is greater than 1 approaches the x-axis when x increases in the negative direction (towards the left-hand side of the graph); so again the x-axis is an asymptote to the curve of the function $y = ac^x$.

It may be wondered just what practical significance such a concept as this could have in practical situations that a biologist might encounter. An example will be presented in the next section of this chapter, but one situation can be described in a general way here.

Frequently it is desirable to describe growth mathematically. This first necessitates measuring some attribute of growth (e.g. weight or length) of an organism, or group of organisms, at several specified points in time, and plotting the results on a graph in which the x-axis represents time and the y-axis the growth attribute (Fig. 4.7). It is now necessary to select a function whose curve will pass as close to the observed points as possible. As discussed in Chapter 3, this requires us first to select a suitable function type and then to estimate the constants of this function from the experimental data. It is the first of these two problems, in relation to growth data, that we now discuss.

The growth of most organisms, particularly of animals, is characterized by the ultimate acquisition of a more or less constant maximum size which could, in theory, be maintained indefinitely were it not for the fact that senescence ensues. However, the phenomena of growth and senescence are not usually studied simultaneously, so that a physiologist in studying growth assumes that a maximum size is maintained constant. To describe this kind of situation mathematically, he would require a function whose curve is asymptotic to a horizontal line $y = a$ (Fig. 4.7) where a is the maximum size attained by the organism concerned. Some realistic 'growth functions' of this kind are described in Chapter 10.

RECTANGULAR HYPERBOLAE

The function y = a/x

It might be thought that a fairly complicated equation would be required to describe a curve with an asymptote. We can see that this is not the case by considering the function $y = a/x$, where a is a constant. Let us simplify further by putting $a = 1$, so that the function is now

$$y = \frac{1}{x} \qquad (4.14)$$

Consider what happens to y in equation 4.14 as x increases in size: to be specific, let x take successively the values 1, 10, 100, 1000 etc., then the corresponding values of y will be 1, 1/10, 1/100, 1/1000, etc., i.e. 1, 0.1, 0.01, 0.001, etc. Evidently, as x increases indefinitely y will get smaller, and will approach, but never actually reach, zero; the line $y = 0$ (i.e. the x-axis) is an asymptote. The x-axis is, however, not the only asymptote; consider what happens to y when x gets very small. Let x take successively the values 1, 0.1, 0.01, 0.001, etc., then y will be correspondingly 1, 10, 100, 1000, etc. In other words, as y increases indefinitely x approaches zero; hence the line $x = 0$ (i.e. the y-axis) is also an asymptote. The curve of $y = 1/x$, resulting from our discussion so far, is shown in Fig. 4.8(a).

Hitherto, we have only considered positive values of x in the equation $y = 1/x$, and these give rise to positive values of y. If x takes negative values, the corresponding values of y will likewise be negative. In fact, the complete curve of the function $y = 1/x$ is as shown in Fig. 4.8(b), and it is interesting to observe that the curve is in two distinct parts. If a straight line is drawn through the origin (broken line in Fig. 4.8(b)) with a gradient of −1, the two parts of the curve are mirror images of one another on either side of this line. The curve of the function $y = a/x$ has a similar appearance to the foregoing curve for any positive value of the constant a; the value of a merely places the curve closer to or further away from the x-axis for a given value of x. If, however, a were negative, the two parts of the curve would be in the other two quadrants formed by the co-ordinate axes.

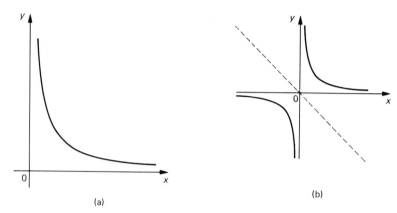

Fig. 4.8   Curves of the rectangular hyperbola $y = 1/x$: (a) portion of the curve in the upper-light hand quadrant defined by the co-ordinate axes, (b) the complete curve.

The conclusion to be drawn from the above discussion is that the x- and y-axes are both asymptotes to the curve of $y = a/x$. Because the two asymptotes are at right-angles to one another, and because the function is a member of a family of functions known as **hyperbolae**, the function $y = a/x$ is called a **rectangular hyperbola**.

The equation $y = a/x$ can be written in the form $y = a(1/x)$. Now if $1/x$ is plotted along the horizontal axis instead of x, then the relationship between y and $1/x$ is linear (Fig. 4.9(a) & (b)). This is a similar result to

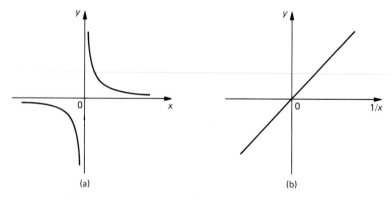

Fig. 4.9   (a) Curve of the rectangular hyperbola $y = 1/x$. (b) The straight line obtained when y is plotted against $1/x$.

that encountered with the function $y = ac^x$, where it was found that plotting log y against x gave a linear relationship. In the present case, the function $y = a(1/x)$ represents a straight line with gradient a and intercept zero.

The relationship between photosynthetic rate and light intensity

The relationship between the gross photosynthetic rate (p) of a leaf, which can be represented on the vertical axis of a graph, and light intensity, I, represented on the horizontal axis, has been found experimentally to be of the form shown by the curve in Fig. 4.10. To quantify the response of photosynthetic rate to light intensity, it is necessary to find a mathematical function which produces such a curve, and one that has been used extensively is

$$p = \frac{1}{a + b/I} \tag{4.15}$$

There are two constants, a and b, and these will take particular values depending on species and the levels of the other external factors. This function is again a rectangular hyperbola, but is not as simple as the one already considered. However, it can be shown that the curve has two asymptotes at right angles to one another: a horizontal one at $p = 1/a$, and a vertical

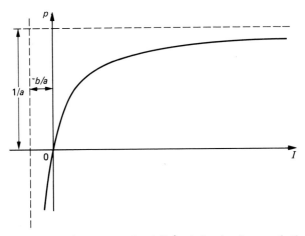

Fig. 4.10   Curve of the function $p = (a + b/I)^{-1}$ relating the photosynthetic rate of a leaf, p, to light intensity, I.

asymptote at $I = -b/a$ (see Chapter 5, page 76 for the method of calculation). The latter asymptote is of no biological interest, since it occurs at a negative light intensity; but the asymptote $p = 1/a$ represents the maximum attainable photosynthetic rate as far as light intensity is concerned (commonly known as the light-saturated photosynthetic rate), under the combination of other factors employed in the experiment.

Equation 4.15 shows that, if light is not limiting, then the rate of photosynthesis is given by $1/a$. Evidently, the lower the value of a, the higher will be the maximum rate of photosynthesis, and *vice versa*. Can we give a physical meaning to a?

One way of interpreting the rate of photosynthesis is to consider the rate of movement of carbon dioxide molecules from the air external to the leaf to the place where they are incorporated into the photosynthetic reduction cycle within the chloroplast. A molecule of CO_2 can be imagined to move along a 'path', from air to carboxylation site, and encounter a number of resistances to its movement along the way. First, the molecule moves through the layer of still air adjacent to the leaf surface to a point immediately over the stomatal cavity. This layer of air is known as the boundary layer, and the resistance to molecular movement through the boundary layer is given the symbol r_a. Then the CO_2 molecule passes through the stoma, encountering another resistance, r_s, the stomatal resistance. The molecule now passes from the sub-stomatal cavity, through the air spaces between the leaf mesophyll cells, then through the membranes of a mesophyll cell and the chloroplast to the site of incorporation into the CO_2 reduction cycle. The resistances encountered here are amalgamated to form the so-called mesophyll resistance, r_m. These three resistances are diffusion resistances, but the fourth, called the carboxylation resistance, r_c, is a biochemical one, and controls the rate of entry of the CO_2 molecule into the reduction cycle. Hence there are four resistances in series, and so the total resistance to the passage of the CO_2 molecule is the sum of these. Now since the constant a in equation 4.15 is inversely related to the maximum photosynthetic rate, it can be identified with the total resistance to the movement of a carbon dioxide molecule from external air to carboxylation site. Thus

$$a = r_a + r_s + r_m + r_c$$

By a different argument, it can be shown that the constant b in equation 4.15 can also have a physical meaning; it represents the quantum efficiency at low light intensities. Hence the utility of data obtained from experiments which involve the measurement of photosynthetic rate, at differing light intensities, is increased by fitting equation 4.15 to such data, and then making comparisons of the constants obtained between species or within the same species under different regimes of other environmental factors.

EXERCISES 4

1. Evaluate (a) $f(0)$ (b) $f(2)$ (c) $f(-2)$, for
$$f(x) = 2x^3 - 3x^2 + 4x + 1$$

2. Evaluate (a) $f(4, 1)$ (b) $f(-1, -2)$, for
$$f(x, y) = 2x^2 + 3xy - y^2 - 3x - 2y + 5$$

3. Draw a graph of the function $p = (3 + 0.5/I)^{-1}$ in the range $0 \leqslant I \leqslant 1.2$. The function represents the relationship between gross photosynthetic rate, p (mg CO_2 absorbed cm^{-2} hr^{-1}), and radiation intensity, I (cal cm^{-2} min^{-1}) in the leaf of a certain plant species. What is the photosynthetic rate when I is:

(a) 0.1   and   (b) 1.0 cal cm^{-2} min^{-1}?

5

Differentiation

This chapter and the three succeeding ones will introduce the branch of mathematics known as the calculus. School mathematics is often rigidly divided into arithmetic, algebra, geometry, and trigonometry, but as the study of mathematics is taken to deeper levels, these somewhat artificial divisions are steadily eroded. These sub-divisions are fairly natural for basic mathematics, but the calculus does not rank with them. The calculus is essentially a tool – a set of mathematical methods invented by Sir Isaac Newton. Problems involving the use of calculus *could* be solved by other methods, but with much greater difficulty and labour.

Calculus deals basically with rates of change, and the facility with which it does this means that it is eminently suited to dealing with dynamic situations as opposed to static ones. All living material is continually in a state of change, and so it is clear that the calculus will be an indispensable tool in biomathematics: it will aid in formulating and solving many problems, such as those concerning growth, biochemical changes, and population dynamics.

The method of the calculus is to discover properties of complete entities by examining the behaviour of their elementary components. An example of the latter, in physics, is in studying the mode of vibration of a stretched string, which we do by considering the motion of each particle of the string independently and then combining all these motions together. This may sound an extremely tedious way to study the vibration of a string, as indeed it would be without the invention of the calculus. By calculus methods, however, the problem can be formulated and solved in a few lines.

Calculus as a subject contains two subdivisions: the differential calculus and the integral calculus. The former is concerned with calculating the rate of change of a function with respect to its variable, e.g. the rate of change of $f(x)$ with respect to x. The latter subdivision of calculus is concerned with the reverse process: given the rate of change of a function with respect to its variable, what is the function itself? The integral calculus also provides techniques for adding together the results obtained for individual components of an entity when we are considering the behaviour of the complete entity.

In this chapter the principles of differentiation, and methods for differentiating mathematical functions are described; but first, a more general mathematical concept must be introduced.

LIMITS

Limits involving infinity

To approach the idea of a mathematical limit, consider again the subject of asymptotes, introduced in the previous chapter. There, we examined the behaviour of the function $y = 1/x$ through a study of its curve, and found that as x increased indefinitely, y approached nearer and nearer to zero. But however large x might be, y was never actually equal to 0.

Let us for a moment imagine that there could be an extremely large number, called infinity, which is given the symbol ∞. It is then intuitively evident that $y = 1/\infty = 0$. That there cannot be an extremely large number called ∞ is obvious from the fact that however large a number one can specify, a still larger number can always be named. Numbers continue increasing without end.

What we can say, however, is that as x increases, $1/x$ decreases, and if x *could* reach infinity $1/x$ *would* be zero. In less imaginary terms, we say that the **limit** of $1/x$, as x approaches infinity, is zero; and the statement is written in symbols as

$$\lim_{x \to \infty} \left(\frac{1}{x} \right) = 0 \qquad (5.1)$$

Another way of viewing the limit can also be seen in this example. As $x \to \infty$, $1/x$ is always positive but decreasing in magnitude; $1/x \to 0$, but never goes below 0, i.e. never starts acquiring negative values. We have already seen this (Fig. 4.8(a)) when considering the curve of $y = 1/x$; this curve approaches the x-axis (i.e. the line $y = 0$) as x increases, but never intersects it. Such a line is an asymptote, and using the definition of an asymptote (page 67) we infer that the line $y = 0$ will 'touch the curve at infinity'.

EXAMPLE 5.1

Find the asymptotes to the curve of the function describing the response of photosynthetic rate, y, to light intensity, x

$$y = \frac{1}{a + (b/x)} \tag{5.2}$$

This question implies finding the limit of y as x approaches ∞, and the limit of x as y approaches ∞. In the first case let $x \to \infty$ in the above equation. Evidently $(b/x) \to 0$, and so $y \to 1/a$. Hence

$$\lim_{x \to \infty} \left(\frac{1}{a + (b/x)} \right) = \frac{1}{a} \tag{5.3}$$

To find the other limit, we first obtain x in terms of y. From 5.2 by cross-multiplication:

$$a + \frac{b}{x} = \frac{1}{y}$$

or

$$\frac{b}{x} = \frac{1}{y} - a$$

and so

$$x = \frac{b}{(1/y) - a}$$

As $y \to \infty$, $1/y \to 0$ and so $x \to -b/a$. Hence we may write

$$\lim_{y \to \infty} \left(\frac{b}{(1/y) - a} \right) = -\frac{b}{a} \tag{5.4}$$

Thus the asymptotes to the curve of equation (5.2) are the horizontal line $y = 1/a$ and the vertical line $x = -b/a$.

Limits not involving infinity

Limits of functions can be obtained when the variable approaches values other than infinity, as the following examples show.

EXAMPLE 5.2

Find the limit of $(3 - x)/(2 + x)$ as

(a) $x \to 2$,   (b) $x \to 3$,   (c) $x \to -2$

(a) Examining the effect of letting x approach 2 in the numerator and denominator separately, we see that as $x \to 2$, $(3 - x) \to 1$ and $(2 + x) \to 4$. Thus

$$\lim_{x \to 2} \left(\frac{3 - x}{2 + x} \right) = \frac{1}{4}$$

(b) Adopting the same approach as in (a), we see that as $x \to 3$, $(3 - x) \to 0$ and $(2 + x) \to 5$. Hence

$$\lim_{x \to 3} \left(\frac{3 - x}{2 + x} \right) = 0$$

(c) Here, as $x \to -2$, $(3 - x) \to 5$ and $(2 + x) \to 0$; and so

$$\lim_{x \to -2} \left(\frac{3 - x}{2 + x} \right) = \infty$$

What this result means in words is that the limit is an infinitely large number.

In Example 5.2(a) and (b), the limits are easy to evaluate. This is because, rather than saying what happens to the expression $(3 - x)/(2 + x)$ when $x \to 2$ or $x \to 3$, we can see the result of putting $x = 2$ and $x = 3$. The value of $(3 - x)/(2 + x)$ when $x = 2$ is $1/4$, and when $x = 3$ the value of the function is $0/5 = 0$. Things are not quite so straightforward in the case of Example 5.2(c) since, by putting $x = -2$, $(3 - x)/(2 + x) = 5/0$ which is indeterminate. However, remembering that a quotient with a very small number in the denominator yields a large number, then we can say that as the denominator *approaches* zero, the quotient itself must get very large. So, when the denominator becomes *vanishingly* small, the whole fraction becomes infinitely large.

EXAMPLE 5.3

Find the limits of

(a) $\dfrac{x^2 - 2x}{x}$ as $x \to 0$

(b) $\dfrac{x^2 - 1}{x - 1}$ as $x \to 1$

(a) If x is put equal to 0, then $(x^2 - 2x)/x = 0/0$, which is completely indeterminate. Since both numerator and denominator are zero in these circumstances, it is not possible to proceed as in Example 5.2(c).

One way of proceeding is as follows. The square of a very small number is smaller than the number itself. So if x is very small, x^2 is negligible compared with x, and so the fraction reduces to $-2x/x = -2$. The smaller x is, the more accurate does the approximation (i.e. ignoring x^2) become; so that when x is vanishingly small, we may write

$$\lim_{x \to 0} \left(\frac{x^2 - 2x}{x} \right) = -2$$

A much better method is to divide out the fraction first; then we have, directly

$$\lim_{x \to 0} \left(\frac{x^2 - 2x}{x} \right) = \lim_{x \to 0} (x - 2) = -2$$

(b) We note that $x^2 - 1 = (x - 1)(x + 1)$. So

$$\lim_{x \to 1} \left(\frac{x^2 - 1}{x - 1} \right) = \lim_{x \to 1} \left\{ \frac{(x - 1)(x + 1)}{x - 1} \right\} = \lim_{x \to 1} (x + 1) = 2$$

Examples 5.3(a) and (b) are particularly relevant to the underlying principle of the differential calculus. It will be observed in each of these two examples that *although the numerator and denominator of the fractions separately tend to zero as x tends to the value shown, the entire quotient remains finite and tends to a finite value.*

It must be emphasised that this very brief introduction to limits has been to aid in understanding the principle of differentiation, rather than as an end in itself. The subject of limits in mathematics is a large and often difficult one, and we have approached it in a rather empirical fashion.

THE PRINCIPLE OF DIFFERENTIATION

Consider the graph of any function, $y = f(x)$ (Fig. 5.1), and let P(x, y) be any point on the curve. Let a point Q be situated close to P (the distance is highly magnified in Fig. 5.1 for clarity) such that the horizontal distance between P and Q is the interval δx, and the vertical distance between the two points is the interval δy. Thus the co-ordinates of point Q are $(x + \delta x, y + \delta y)$.

On a point of symbolization, an interval (that is, a distance between points), is usually denoted by the capital Greek letter delta, Δ, but if the interval is small the equivalent lower case (small) letter is employed, δ. Thus an interval on an x-scale is Δx, but if the interval is small it is denoted by δx. Note that neither Δ nor δ are algebraic quantities, but are a kind of prefix. In the term Δx, Δ and x can never be separated.

Reverting to Fig. 5.1, the straight line PQ is known as a **chord** of the curve, and its gradient is obviously $\delta y / \delta x$. The point made in the previous paragraph, that δ is not an algebraic quantity but a prefix, applies here. There is no question of the δ's cancelling out; the term $\delta y / \delta x$ means a small interval in y divided by a small interval in x.

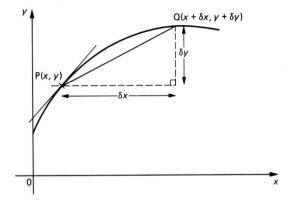

Fig. 5.1 The principle of differentiation (see page 78).

Now let point Q move along the curve towards point P. When this happens δx (and δy), already small, decrease in size — approach zero in fact; but although δx and δy both approach zero, their quotient remains finite (c.f. page 78). Evidently the gradient of PQ will change under these circumstances and the chord PQ will, in fact, approach the tangent at P. Hence, when δx and δy become vanishingly small, *but not zero*, the chord PQ will be indistinguishable from the tangent at point P; more important, the *gradient* of chord PQ will be indistinguishable from the *gradient* of the tangent at P under these conditions.

The gradient of the tangent at P is the limit of the gradient of the chord PQ as PQ becomes vanishingly small, and is denoted by dy/dx; so

$$\lim_{\delta x \to 0} \left(\frac{\delta y}{\delta x} \right) = \frac{dy}{dx} \tag{5.5}$$

In 5.5, $\delta y/\delta x$ is the gradient of the chord and dy/dx is the gradient of the tangent as already defined; so 5.5 can be expressed in words as, 'the limit of the gradient of a chord, $\delta y/\delta x$, as δx tends to zero, is the gradient of a tangent, dy/dx'. In the term dy/dx, the d's are again prefixes, denoting vanishingly small intervals.

Differentiation from first principles

To fix our ideas, let us differentiate some of the functions we encountered in the previous chapter.

EXAMPLE 5.4

Differentiate the following functions with respect to x.

(a) $y = x^3$,   (b) $y = 1/x$,   (c)* $y = ac^x$

(a) This is a very simple cubic polynomial containing only a term in x^3 with a coefficient of 1. The curve is shown in Fig. 5.2, and P is any point on the curve with co-ordinates (x, y). Clearly there is a relationship between x and y, the functional relationship

$$y = x^3 \tag{5.6}$$

Consider a point Q, which is reached by moving a small distance along the curve from P so that its position is δx from P in the direction of the x-axis, and a corresponding small increment (δy) of y from P. Point Q will have co-ordinates $(x + \delta x, y + \delta y)$, and again there is a relationship between the two co-ordinates:

$$y + \delta y = (x + \delta x)^3 \tag{5.7}$$

Now subtract 5.6 from 5.7

$$(y + \delta y) - y = (x + \delta x)^3 - x^3$$

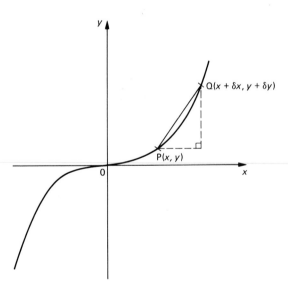

Fig. 5.2   Application of the principle of differentiation to the function $y = x^3$ (see Example 5.3(a)).

i.e.

$$\delta y = (x + \delta x)^3 - x^3$$

Next, expand the cubed bracket, so that we have

$$\delta y = x^3 + 3x^2 \cdot \delta x + 3x(\delta x)^2 + (\delta x)^3 - x^3$$

in which the two x^3 terms cancel one another out, and we are left with

$$\delta y = 3x^2 \cdot \delta x + 3x(\delta x)^2 + (\delta x)^3$$

On dividing both sides of this expression by δx, we have

$$\frac{\delta y}{\delta x} = \frac{3x^2 \cdot \delta x}{\delta x} + \frac{3x(\delta x)^2}{\delta x} + \frac{(\delta x)^3}{\delta x}$$

and, on cancelling δx's on the right-hand side:

$$\frac{\delta y}{\delta x} = 3x^2 + 3x \cdot \delta x + (\delta x)^2 \tag{5.8}$$

Equation 5.8 is an expression by means of which the gradient of chord PQ can be evaluated for a given x and δx.

Now let δx tend to zero. This means that δx becomes so small that it can be taken as zero in expression 5.8, so that the second and third terms on the right-hand side of that equation disappear, leaving only $3x^2$. The quotient $\delta y/\delta x$ on the left-hand side thus remains finite and tends to a finite limit, $3x^2$, which we again denote as dy/dx (as in equation (5.5)). So, in the present example

$$\lim_{\delta x \to 0} \{3x^2 + 3x \cdot \delta x + (\delta x)^2\} = 3x^2 \tag{5.9}$$

On comparing 5.5 and 5.9, we see that $dy/dx = 3x^2$. Hence, for the function $y = x^3$, $dy/dx = 3x^2$. This means that the gradient of the tangent to the curve $y = x^3$, at a point whose x-co-ordinate is x, is given by $3x^2$.

(b) This is the simplest rectangular hyperbola (page 69), and one part of the curve is reproduced in Fig. 5.3. The differentiation procedure is exactly the same as in the previous example, so only the outline will be given here. At point P, we have the relationship

$$y = \frac{1}{x} \tag{5.10}$$

At point Q, we have the similar relationship

$$(y + \delta y) = \frac{1}{(x + \delta x)} \tag{5.11}$$

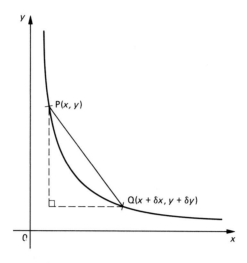

Fig. 5.3 Application of the principle of differentiation to the function $y = 1/x$ (see Example 5.4 (b)).

5.11 minus 5.10 gives

$$\delta y = \frac{1}{(x + \delta x)} - \frac{1}{x}$$

Put the right-hand side over a common denominator:

$$\delta y = \frac{x - (x + \delta x)}{x(x + \delta x)} = \frac{-\delta x}{x(x + \delta x)}$$

Dividing both sides by δx gives

$$\frac{\delta y}{\delta x} = \frac{-1}{x(x + \delta x)} \qquad (5.12)$$

As before, equation 5.12 gives the gradient of the chord PQ for a given x and δx. Hence it is evident that

$$\lim_{\delta x \to 0} \left\{ -\frac{1}{x(x + \delta x)} \right\} = -\frac{1}{x^2} \qquad (5.13)$$

Again, comparing 5.5 and 5.13, we see that $dy/dx = -1/x^2$.

(c) This example is best omitted at a first reading, as it uses a result to be given later in the book. The example is put here to show that all the functions we have so far considered can be differentiated from first principles.

At a point P, whose co-ordinates are (x, y), we have the relationship

$$y = ac^x \qquad (5.14)$$

Since the variable, x, is now in the exponent of the function, it is easiest to start by taking logarithms of both sides; and as usual in mathematical work it is best to use natural logarithms. Thus

$$\log_e y = \log_e a + x \cdot \log_e c \qquad (5.15)$$

At point Q, whose co-ordinates are $(x + \delta x, y + \delta y)$, we have the relationship

$$\log_e(y + \delta y) = \log_e a + (x + \delta x) \cdot \log_e c. \qquad (5.16)$$

Subtracting 5.15 from 5.16 gives

$$\log_e(y + \delta y) - \log_e y = (x + \delta x) \cdot \log_e c - x \cdot \log_e c$$

Remembering that the logarithm of the quotient of two quantities is equal to the difference of the logarithms of the two quantities, the above expression reduces to

$$\log_e\left(\frac{y + \delta y}{y}\right) = (x + \delta x) \cdot \log_e c - x \cdot \log_e c$$

Expand the right-hand side, and divide out on the left-hand side.

$$\log_e\left(1 + \frac{\delta y}{y}\right) = x \cdot \log_e c + \delta x \cdot \log_e c - x \cdot \log_e c$$

i.e.

$$\log_e\left(1 + \frac{\delta y}{y}\right) = \delta x \cdot \log_e c$$

Now, since $\delta y/y$ is small

$$\log_e\left(1 + \frac{\delta y}{y}\right) \simeq \frac{\delta y}{y} \qquad \text{(page 174)}$$

So, we have

$$\frac{\delta y}{y} \simeq \delta x \cdot \log_e c$$

Divide both sides by δx and cross-multiply the y in the denominator of the left-hand side up to the right-hand side.

$$\frac{\delta y}{\delta x} \simeq y \cdot \log_e c$$

From 5.14 we know that $y = ac^x$, so substituting for y:

$$\frac{\delta y}{\delta x} \simeq ac^x \cdot \log_e c \tag{5.17}$$

Equation 5.17 becomes more and more accurate as δx and δy decrease in size; hence, invoking the limiting relationship 5.5, we may write

$$\frac{dy}{dx} = ac^x \cdot \log_e c$$

DIFFERENTIATION OF POLYNOMIAL FUNCTIONS

The rule for differentiating y = ax^n

In the first example above, we found that if $y = x^3$, then $dy/dx = 3x^2$, i.e. $3x^{(3-1)}$. In the second example, $y = 1/x$ and $dy/dx = -1/x^2$. Now $1/x$ can be written as x^{-1}, and $-1/x^2$ as $-x^{-2}$; thus we have that when $y = x^{-1}$, $dy/dx = -x^{-2}$, i.e. $-1x^{(-1-1)}$. We see that in both of these examples we are differentiating an expression of the form $y = x^n$, where $n = 3$ in the first case, and $n = -1$ in the second. In both instances, the result for dy/dx is obtained if x is multiplied by n, and 1 is deducted from n in the exponent. Thus, if

$$y = x^n,$$

then

$$\frac{dy}{dx} = nx^{(n-1)}$$

$$\left.\right\} \tag{5.18}$$

The two expressions in 5.18 state a general rule, which applies if n is any positive or negative integer.

Now suppose that x^n is multiplied by a constant a (i.e. by a coefficient), giving ax^n. It can easily be demonstrated (Exercise 1 at the end of this chapter) that if $y = ax^n$, $dy/dx = anx^{(n-1)}$. In other words, the coefficient, a, remains unchanged, and does not affect the operation of the basic rule.

EXAMPLE 5.5
Differentiate the following functions with respect to x, by the rule provided above.

(a) $y = 5x^2$ (b) $y = -3x^5$ (c) $y = \dfrac{2}{x^2}$

(a) We can straight away write

$$\frac{dy}{dx} = (5)2x^{(2-1)} = 10x$$

(b) Again, we have $dy/dx = (-3)5x^{(5-1)} = -15x^4$

(c) Write the expression as $y = 2x^{-2}$.

Then

$$\frac{dy}{dx} = (2)(-2)x^{(-2-1)} = -4x^{-3}$$

i.e.

$$\frac{dy}{dx} = -\frac{4}{x^3}$$

Differentiation of roots

Although it will not be proved here, the rule in equations 5.18 also applies when the exponent, n, is a fraction. Recalling that a fractional exponent implies a root (page 13), e.g. $x^{1/2} = \sqrt{x}$, this means that we can use the rule to differentiate roots of x as well as x raised to powers.

EXAMPLE 5.6

Differentiate the following expressions with respect to x.

(a) $y = \sqrt[4]{x}$ (b) $y = 3\sqrt{x}$ (c) $y = 1/\sqrt[3]{x}$ (d) $y = \sqrt{(x^3)}$

(a) Write the expression as $y = x^{1/4}$, then

$$\frac{dy}{dx} = \frac{1}{4}x^{(1/4-1)}$$

i.e.

$$\frac{dy}{dx} = \frac{1}{4}x^{-3/4} = \frac{1}{4x^{3/4}}$$

Thus

$$\frac{dy}{dx} = \frac{1}{4\sqrt[4]{(x^3)}}$$

(b) $y = 3x^{1/2}$,   then $\frac{dy}{dx} = 3\left(\frac{1}{2}\right)x^{(1/2-1)} = \frac{3}{2}x^{-1/2}$

i.e.

$$\frac{dy}{dx} = -\frac{3}{2x^{1/2}} = -\frac{3}{2\sqrt{x}}$$

(c) $y = 1/x^{1/3}$,   i.e. $y = x^{-1/3}$

Hence

$$dy/dx = (-1/3)x^{-4/3} = -1/(3x^{4/3})$$

Thus

$$dy/dx = -1/\{3\sqrt[3]{(x^4)}\}$$

(d) $y = x^{3/2}$,   hence $dy/dx = (3/2)x^{1/2} = 3\sqrt{x}/2$

Differentiation of a constant

The result is most easily demonstrated by means of an example.

EXAMPLE 5.7
Differentiate $y = a$ (where a is constant), with respect to x. It is implied that y is a function of x even though x does not actually appear in the function; however, the expression could be written as $y = ax^0$, which now clearly demonstrates that y can be regarded as a function of x. This means that we can draw a graph of the function $y = a$ on an x–y plane, and this is shown in Fig. 5.4. The 'curve' of $y = a$ is a horizontal straight line, and

Fig. 5.4   The straight line $y = a$.

we can differentiate from first principles in the usual way by considering two points on the line, $P(x, y)$ and $Q(x + \delta x, y + \delta y)$. For P, $y = a$; and for Q, $y + \delta y = a$. Subtracting, we have $\delta y = 0$, and so $\delta y/\delta x = dy/dx = 0$. So, **differentiating a constant yields zero**. This is to be expected, since the gradient of a horizontal line is zero.

Differentiation of a polynomial function

A polynomial function of x consists of a series of terms of the form ax^n added together. Can we differentiate a polynomial by dealing with each term separately? Example 5.8 shows that we can.

EXAMPLE 5.8

Differentiate $y = x^2 - 5x + 2$ with respect to x from first principles.

In the usual way, formulate two equations for points P and Q which lie on the curve of the function.

$$y = x^2 - 5x + 2 \tag{5.19}$$

$$y + \delta y = (x + \delta x)^2 - 5(x + \delta x) + 2 \tag{5.20}$$

Subtraction of 5.19 from 5.20 gives

$$\delta y = (x + \delta x)^2 - 5(x + \delta x) - x^2 + 5x$$

Expand the brackets

$$\delta y = x^2 + 2x \cdot \delta x + (\delta x)^2 - 5x - 5 \cdot \delta x - x^2 + 5x$$

i.e.

$$\delta y = 2x \cdot \delta x + (\delta x)^2 - 5 \cdot \delta x$$

Divide both sides by δx

$$\frac{\delta y}{\delta x} = 2x + \delta x - 5$$

and so

$$\lim_{\delta x \to 0} (2x + \delta x - 5) = 2x - 5 = \frac{dy}{dx}$$

If 5.19 is differentiated term by term using the rule in 5.18, we obtain the same result as in Example 5.8. Hence, a polynomial function can be differentiated by applying the rule for differentiating ax^n to each term in the polynomial separately.

THE DIFFERENTIAL COEFFICIENT

We have developed the discussion of differentiation by depicting the various mathematical activities on graphs; thus, we have given dy/dx a meaning: it is the gradient of the tangent to a curve, and hence the gradient of the curve itself at that point (see page 79). The differentiation of a function can, however, be accomplished without any reference to a curve, and without giving the result a physical meaning. It is not even necessary to invoke a second variable, e.g. y; we can work entirely in terms of the one variable x. For example, we have already seen that if $y = x^n$, then $dy/dx = nx^{(n-1)}$; but we could embody both these expressions into one, and write

$$\frac{d(x^n)}{dx} = nx^{(n-1)} \tag{5.21}$$

Instead of y in the numerator of the left-hand side, we have written what y is in terms of x, namely x^n. So Example 5.8 could have been stated, 'Differentiate $x^2 - 5x + 2$ with respect to x from first principles', or more succinctly, 'Find $d(x^2 - 5x + 2)/dx$ from first principles'.

Reverting to two variables for a moment, consider the case where y is an unknown function of x, and we write $y = f(x)$. Then the differentiated function (which is also unknown) is written as $dy/dx = f'(x)$. This notation can also be used when the second variable, y, is not employed; e.g. if $f(x) = x^2 - 5x + 2$, then $f'(x) = 2x - 5$.

If a function of x, $f(x)$, is differentiated, we obtain a new function of x which we write as $f'(x)$. The function $f'(x)$ is *derived* from the original function, $f(x)$, by the process of differentiation, and $f'(x)$ may be called the **derivative** of $f(x)$ but is also known as the **differential coefficient**. Thus, for example, $3x^2$ is the differential coefficient (or derivative) of x^3.

DIFFERENTIATION OF MORE COMPLICATED FUNCTIONS

Differentiation of a sum or difference of terms

So far, we have differentiated from first principles most of those functions that were introduced in the previous chapter, namely, the polynomials, the simple rectangular hyperbola, and functions of the form $y = ac^x$.

In the case of polynomials, it was found that each term could be differentiated separately. As an example for further discussion, consider the polynomial function $y = x^2 - 5x + 2$. Now the term 'function' can be used at many different levels of complexity. For instance, in this example if we let $y = f(x)$, then $f(x) = x^2 - 5x + 2$. But each of the terms on the right-hand side are also functions of x; even the last term is a function of x as it could be written as $2x^0$. Thus, we could put $F(x) = x^2$, $\phi(x) = -5x$, and $\psi(x) = 2$; and so

$$f(x) = F(x) + \phi(x) + \psi(x) \qquad (5.22)$$

In other words, a function of x that is a polynomial is made up of a number of simple functions of x added together.

We have already seen that $f'(x) = 2x - 5$ for this example (page 87). Differentiating the individual functions on the right-hand side of 5.22, we find $F'(x) = 2x$, $\phi'(x) = -5$, and $\psi'(x) = 0$. So again we have the result that $f'(x) = 2x - 5$. This means that we can formulate the following rule.

The differential coefficient of a function which consists of a number of terms added together is equal to the sum of the differential coefficients of each term.

The term 'addition' also includes 'subtraction'. In symbols, for just two terms, if

then
$$\left. \begin{array}{l} f(x) = \phi(x) \pm \psi(x) \\[2mm] f'(x) = \phi'(x) \pm \psi'(x) \end{array} \right\} \qquad (5.23)$$

Differentiation of a product

Suppose we have the following function of x to differentiate: $f(x) = (x^3 - 2)(x + 1)$. One way of doing it would be to multiply out the brackets first, which would give a polynomial, and then differentiate the latter in the usual way. This example as originally given, however, is obviously made up of two functions of x multiplied together. If we let $\phi(x) = x^3 - 2$ and $\psi(x) = x + 1$, then $f(x) = \phi(x) \cdot \psi(x)$. Now, just as we found above that there is a rule for differentiating a sum or a difference of two functions of x, so there is a rule to differentiate a product of two functions of x, and it is often much more convenient to apply than it is to multiply out the given product first. The rule, which we shall not prove here, is rather cumbersome in words; so in symbols, if

then
$$\left. \begin{array}{l} f(x) = \phi(x) \cdot \psi(x) \\[2mm] f'(x) = \phi(x) \cdot \psi'(x) + \psi(x) \cdot \phi'(x) \end{array} \right\} \qquad (5.24)$$

EXAMPLE 5.9

Find

$$d\{(x^3 - 2)(x + 1)\}/dx$$

In the notation of 5.24, we have

$$\phi(x) = x^3 - 2 \quad \text{and} \quad \psi(x) = x + 1,$$

so

$$\phi'(x) = 3x^2 \quad \text{and} \quad \psi'(x) = 1.$$

Hence

$$f'(x) = (x^3 - 2)(1) + (x + 1)(3x^2),$$

i.e.

$$f'(x) = 4x^3 + 3x^2 - 2.$$

We can check this result by multiplying out the brackets first:

$$f(x) = x^4 + x^3 - 2x - 2, \quad \text{and so} \quad f'(x) = 4x^3 + 3x^2 - 2.$$

Differentiation of a quotient

In principle, the rule for the differentiation of a quotient is similar to that for a product. If

$$\left. \begin{array}{l} f(x) = \phi(x)/\psi(x) \\ f'(x) = \dfrac{\psi(x) \cdot \phi'(x) - \phi(x) \cdot \psi'(x)}{\{\psi(x)\}^2} \end{array} \right\} \qquad (5.25)$$

EXAMPLE 5.10

Find

$$d\{(x^3 - 2)/(x + 1)\}/dx$$

$$\phi(x) = x^3 - 2 \quad \text{and} \quad \psi(x) = x + 1.$$

Hence

$$\phi'(x) = 3x^2 \quad \text{and} \quad \psi'(x) = 1;$$

then

$$f'(x) = \frac{3x^2(x + 1) - (x^3 - 2)}{(x + 1)^2}$$

Function of a function

Suppose we have to differentiate with respect to x the function $(5x^2 + 7x + 2)^6$. There are three ways in which this could be done: (i) expand the bracket and then differentiate the resulting polynomial term by term (excessively laborious, and very liable to error); (ii) write out the bracketed term six times and then use an extension of the product rule (still very laborious); or (iii) use a new rule called the function of a function (very much quicker), as described in the following example.

EXAMPLE 5.11
  Find

$$d\{(5x^2 + 7x + 2)^6\}/dx$$

When using the function of a function rule, it is convenient to start by putting the function to be differentiated equal to another variable, if this is not already specified; e.g. let

$$y = (5x^2 + 7x + 2)^6 \qquad (5.26)$$

The procedure then is to put some suitable part of the function, which can be differentiated by some other method, equal to yet another variable. In this example let

$$u = 5x^2 + 7x + 2 \qquad (5.27)$$

then 5.26 can be written as

$$y = u^6 \qquad (5.28)$$

Now differentiate 5.28 with respect to u:

$$\frac{dy}{du} = 6u^5 \qquad (5.29)$$

and differentiate 5.27 with respect to x:

$$\frac{du}{dx} = 10x + 7 \qquad (5.30)$$

The function of a function rule states that, in the present notation,

$$\frac{dy}{dx} = \frac{dy}{du} \cdot \frac{du}{dx} \qquad (5.31)$$

We see from 5.26 that the answer to the original problem will be given by dy/dx — the left-hand side of 5.31; and it is evident that we already have the two terms on the right-hand side of 5.31, in 5.29 and 5.30. Hence

$$\frac{dy}{dx} = (6u^5)(10x + 7) \qquad (5.32)$$

Expression 5.32 is not, however, quite the final answer. The original function contained only the variable x, and the answer must do so too. So, first eliminating u using 5.27, we have

$$\frac{dy}{dx} = 6(5x^2 + 7x + 2)^5(10x + 7)$$

and so, finally, we have that

$$\frac{d\{(5x^2 + 7x + 2)^6\}}{dx} = 6(5x^2 + 7x + 2)^5(10x + 7)$$

EXAMPLE 5.12
Differentiate the following functions with respect to x.

(a) $1/\sqrt{(2x^3 + 4)}$ (b) $x \cdot \sqrt{(2x + 1)}$

(a) At first sight, it might be thought that the quotient rule would be appropriate here. But when differentiation of the denominator is attempted, the function of a function rule has to be used. It is more convenient to use the function of a function rule only, since the given function can be written as $(2x^3 + 4)^{-1/2}$ using the laws of indices. So, let

$$y = (2x^3 + 4)^{-1/2}$$

and let

$$u = 2x^3 + 4$$

giving

$$y = u^{-1/2}$$

Now

$$\frac{dy}{du} = -\frac{1}{2}u^{-3/2}$$

and

$$\frac{du}{dx} = 6x^2$$

Thus

$$\frac{dy}{dx} = \left(-\frac{1}{2}u^{-3/2}\right)\left(6x^2\right) = -3x^2u^{-3/2} = -\frac{3x^2}{\sqrt{(u^3)}}$$

and so

$$\frac{d\{1/\sqrt{(2x^3 + 4)}\}}{dx} = -\frac{3x^2}{\sqrt{\{(2x^3 + 4)^3\}}}$$

(b) This example is a product of two functions of x; i.e. x, and $(2x + 1)^{1/2}$. The latter, however, needs to be differentiated by the function of a function rule first. Let

$$y = (2x + 1)^{1/2}$$

and let

$$u = 2x + 1$$

giving

$$y = u^{1/2}$$

Now

$$\frac{dy}{du} = \frac{1}{2}u^{-1/2}$$

and

$$\frac{du}{dx} = 2$$

Thus

$$\frac{dy}{dx} = \left(\frac{1}{2}u^{-1/2}\right)(2) = u^{-1/2}$$

So

$$\frac{d\{\sqrt{(2x + 1)}\}}{dx} = \frac{1}{\sqrt{(2x + 1)}}$$

Now apply the product rule to the whole expression:

$$\frac{d\{x \cdot \sqrt{(2x + 1)}\}}{dx} = \frac{x}{\sqrt{(2x + 1)}} + \sqrt{(2x + 1)} = \frac{3x + 1}{\sqrt{(2x + 1)}}$$

HIGHER ORDER DIFFERENTIAL COEFFICIENTS

Imagine that we have the following quadratic function

$$y = 5x^2 + 7x + 2 \tag{5.33}$$

Differentiation gives

$$\frac{dy}{dx} = 10x + 7 \tag{5.34}$$

Suppose, however, we had been given the function

$$y = 10x + 7; \qquad\qquad (5.35)$$

differentiation of this gives

$$\frac{dy}{dx} = 10. \qquad\qquad (5.36)$$

Now clearly, the actual functions of x involved in the right-hand sides of expressions 5.34 and 5.35 are identical. The only difference is that 5.34 represents the differential coefficient of 5.33, while 5.35 has not been written here as a differential coefficient. But obviously the function $10x + 7$ can be differentiated with respect to x, and yields 10. Is there any reason why we should not say that differentiation of 5.33 gives 5.34, and that differentiation of 5.34 gives 5.36? No reason at all: in fact $5x^2 + 7x + 2$ differentiated gives $10x + 7$, and when differentiated again gives 10. In symbols:

$$y = 5x^2 + 7x + 2 \quad \text{or} \quad f(x) = 5x^2 + 7x + 2$$

$$\frac{dy}{dx} = 10x + 7 \qquad\quad \text{or} \quad f'(x) = 10x + 7$$

$$\frac{d^2y}{dx^2} = 10 \qquad\qquad \text{or} \quad f''(x) = 10$$

The equations on the left above give the notations when the situation warrants the use of two variables, y being a function of x, and the equations on the right show the symbols used when we have no desire to mention another variable, y.

It has been stated that dy/dx ($= f'(x)$) is called the derivative, or differential coefficient. This terminology now requires expansion: we say that dy/dx ($= f'(x)$) is the **first derivative**, or the **first differential coefficient**. Likewise, d^2y/dx^2 (which in words is read as, 'd-two-y by d-x-squared', and not as, 'd-squared-y etc.') and $f''(x)$ denote the **second derivative**, or the **second differential coefficient**.

There is, in fact, nothing to stop us obtaining differential coefficients of higher order than the second, but as they are of no practical value to the biologist, we shall leave the subject at this point.

EXERCISES 5

1. By differentiating from first principles, show that

$$\frac{d(ax^4)}{dx} = 4ax^3$$

In 2 & 3, differentiate the functions with respect to x.

2. (a) $(3x - 2)^3$ (b) $1/(ax^2 + bx + c)$ (c) $(x + 1/x)^n$
3. (a) $(1 + x^2)/(1 - x^2)$ (b) $x/\sqrt{(1 - x)}$ (c) $x^2 \cdot \sqrt{(a^2 - x^2)}$
    (d) $\{(ax + b)/(cx + d)\}^n$
4. Find dy/dx and d^2y/dx^2 for the functions
    (a) $y = 1/x$ (b) $y = ac^x$

6

Use of the Differential Calculus

THE PHYSICAL MEANINGS OF THE FIRST AND SECOND DERIVATIVES

Before any mathematical concept can be put to use, the symbols and procedures involved must be identified with quantities and processes which occur in the physical world. So far, we have seen that the first order differential coefficient of a function represents the gradient of the curve of that function, but we have not yet attempted to assign any kind of meaning to the second derivative.

Now a mathematical function $f(x)$ is a completely abstract entity. For any value of x, $f(x)$ can *usually* be evaluated; but this is not always possible: consider, for instance, the function $1/x$ when $x = 0$. Leaving aside these kinds of difficulties, which are rare in practice, we can say that given a numerical value of x, a numerical value of $f(x)$ can be calculated. If this is done for a number of different x-values, we could write the results in a table. Now x and $f(x)$ are obviously variable quantities which we can identify with the axes of a two-dimensional graph, x for the horizontal axis — the independent variable — because it can take any value we please, and $f(x)$ for the vertical axis — the dependent variable — because its values depend on those of x. Often, $f(x)$ is given a single symbol of its own, y, but this is not necessary.

By using the tabulated values of $f(x)$ for each x, a curve may be drawn with respect to the two axes, and this curve is a *physical representation* of the function. Again, the first differential coefficient, $f'(x)$, is a purely abstract concept, derived from the original function, $f(x)$, which is also abstract; but we have already shown that $f'(x)$ represents the gradient of the curve of $f(x)$, which gives the first derivative a physical meaning.

However, the foregoing discussion, although a step in the right direction, is not sufficient for us to be able to use the differential coefficient in order to solve problems arising in the real world. We now need to identify features connected with the curve of a function with happenings in the physical world, and a simple example is to consider the motion of a particle in a straight line along the top of a bench.

Distance and velocity — the first differential coefficient

Let us imagine a marble at rest at zero time ($t = 0$) on one end of the bench (zero distance, $s = 0$). Assume now that the marble starts moving. For this to happen, the marble must, of course, accelerate, but it is not necessary here to enquire into the force that must be applied to the marble to cause it to accelerate.

We shall assume that the initial acceleration decreases linearly with time, and the situation is shown in Fig. 6.1(a). The effect can be likened to a car driver very gradually easing his foot off the accelerator pedal, so that the car still accelerates but at a continuously decreasing rate. If this decrease in acceleration is continued, there comes a time when the marble is undergoing zero acceleration, shown in Fig. 6.1(a) where the straight line intersects the horizontal (time) axis. Thereafter, if the same linear relationship between acceleration and time continues to hold, the acceleration becomes negative, that is, a retardation. Obviously, if the retardation continues long enough, the marble will stop, at time T (say). So to summarize, a marble at rest zero time and zero distance from one edge of the bench starts to move under the influence of a relatively high acceleration. This acceleration declines linearly with time, eventually becoming a retardation until the marble comes to rest again when $t = T$.

Now consider the *distance* the marble travels as a function of time. Evidently while the marble is accelerating, the distance moved in a unit of time will increase, and *vice versa* while the marble is retarding. So the graph of distance moved, s, in time t can be represented by the curve in Fig. 6.1(b). We can assume that the curve shows s as some function of t, i.e. $s = f(t)$, and in order to make further progress, it will be necessary to find a possible form of $f(t)$ which will produce a curve like that in Fig. 6.1(b).

In Fig. 6.2(a) is shown the graph of a third degree polynomial (c.f. Fig. 4.3(c)). The curve is positioned so that: (i) it passes through the origin, (ii) it rises when t has small positive values, and (iii) it ceases to rise beyond a certain value of t (i.e. beyond a certain time) which we call T. Evidently, that part of the curve lying in the range $0 \leqslant t \leqslant T$ will adequately describe the distance that the marble has travelled, as shown in

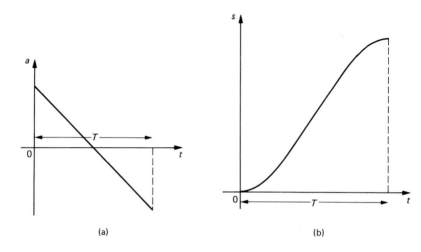

Fig. 6.1   (a) An acceleration—time curve whose equation is $a = 3(T - 2t)$. (b) A portion of the corresponding distance—time curve whose equation is $s = 3Tt^2/2 - t^3$.

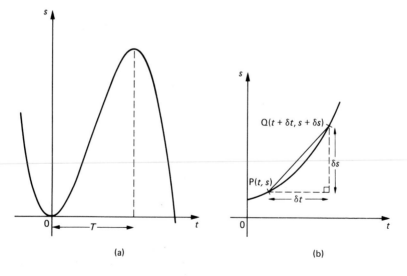

Fig. 6.2   (a) A more complete graph of the curve shown in Fig. 6.1(b). (b) Two points on a distance—time curve with line PQ depicting average velocity between the points.

Fig. 6.1(b). It can be demonstrated that the equation of the curve in
Fig. 6.2(a) is

$$s = \frac{3}{2}Tt^2 - t^3 \qquad (6.1)$$

and if we compare this with the full equation for a third degree polynomial
in the same notation, $s = a + bt + ct^2 + dt^3$, we see that the constant term
and the term in t are missing (i.e. the coefficients a and b are zero, written
$a = b = 0$), $c = 3T/2$, and $d = -1$. On the other hand, equation 6.1 only
describes the motion of our marble when $0 \leqslant t \leqslant T$. Before zero time, i.e.
before the time the marble starts to move, it is considered to be stationary,
$s = 0$; the portion of the curve corresponding to negative values of t does
not describe the motion of the marble. Similarly, when $t > T$, the marble
is at rest distance S from its starting point, whereas the curve of 6.1 shows
the marble retracing its steps! Substituting $T = t$ in 6.1 gives $S = T^3/2$. So
the full quantitative description of the motion of the marble is given thus:

$$\left. \begin{array}{ll} s = \dfrac{3}{2}Tt^2 - t^3 & 0 \leqslant t \leqslant T \\[2mm] s = 0 & t < 0 \\[2mm] s = T^3/2 & t > T \end{array} \right\} \qquad (6.2)$$

Notice that the variable t does not appear in the last two equations of 6.2,
thus showing that s (distance travelled) is independent of time here; i.e. s is
constant.

Now imagine two points situated on the distance-time curve
(Fig. 6.2(b)), P(t, s) giving the distance travelled at time t, and Q($t + \delta t$,
$s + \delta s$) giving the distance travelled at time ($t + \delta t$), a short time later. The
gradient of the chord PQ is, of course, given by $\delta s/\delta t$. This quotient
denotes the amount of distance travelled, δs, in the small interval of time,
δt. For a given δt, the larger that δs is, the greater the distance covered.
Hence, $\delta s/\delta t$ is a measure of the speed or velocity with which the marble
is moving. Moreover, because the marble is undergoing acceleration
(Fig. 6.1(a)), its velocity is not constant; hence $\delta s/\delta t$ is the **average velocity**
over the **time interval δt**.

Now we know that the gradient of the curve at P is the limiting value of
$\delta s/\delta t$ as $\delta t \to 0$; we can write this as ds/dt. So as we have identified the
gradient of the chord PQ with the average velocity over the time interval
δt, then the gradient of the curve (i.e. the gradient of the tangent) at point
P must represent a velocity also. Since the time interval is now vanishingly
small, there is no question of an average velocity over an interval of time,
and the velocity measured by the gradient of the curve at time t is the
instantaneous velocity of the marble **at time t**.

The idea of an instantaneous velocity may be rather hard to grasp, so consider a slightly different example. When a train is said to be moving at 60 miles per hour, this does not mean that it will travel 60 miles in the next hour, or that it has travelled 60 miles in the previous hour, since its speed will fluctuate. What the statement means is that, 'at the *present time* the train is travelling at 60 miles per hour.' How do we define 'the present time'? The speed of 60 miles per hour is equivalent to one mile per minute, or, more usefully, to 88 feet per second; we can say that the train is much more likely to travel 88 feet in the next second, than 60 miles in the next hour, since its speed is much less likely to change in the next second than in the coming hour. It would be even more accurate to say that in the next 1/10th of a second the train would travel 8.8 feet, more accurate still to say that 0.88 feet would be traversed in the next 1/100th of a second, and so on. With a decreasing interval of time, we have a proportionately smaller distance travelled, but nevertheless, the speed of the train is still 60 miles per hour, and the statement that the train is travelling at 60 miles per hour becomes more accurate as the time interval denoted as 'the present time' becomes smaller. When 'the present time' becomes a vanishingly small interval of time, the statement concerning the speed of the train is as accurate as it can be.

Reverting to our marble moving along a bench, we recall that the distance travelled from its starting point, measured as a function of time, is given by $s = 3Tt^2/2 - t^3$. Differentiating,

$$\frac{ds}{dt} = 3Tt - 3t^2 \tag{6.3}$$

or

$$\frac{ds}{dt} = 3t(T - t) \tag{6.4}$$

The differential coefficient is a function of time, and equation 6.3 shows it to be a second degree polynomial, which has been factorised in 6.4. Either equation, 6.3 or 6.4, describes the velocity of the marble when t lies in the range $0 \leqslant t \leqslant T$.

Now when the marble is at rest, velocity is zero, i.e. $ds/dt = 0$. To find the values of t when ds/dt is zero implies solving the quadratic equation

$$3t(T - t) = 0$$

Either $3t = 0$ or $T - t = 0$, i.e. $t = 0$ or $t = T$. This confirms what we already know, that the marble is stationary when $t = 0$ and $t = T$ (Fig. 6.1(b)).

So to summarize, and put into more general terms, if we have a function of x, such as $y = f(x)$, then $dy/dx = f'(x)$ is the instantaneous velocity of y with respect to x, or the **instantaneous rate of change** of y with respect to

x. An **average rate of change** between two specified values of x, x and $x + \delta x$, is given by $\delta y / \delta x$.

Velocity and acceleration — the second differential coefficient

Still using the example of a marble in motion, we have seen that the velocity, for which the symbol v is generally used, is given by ds/dt and bears the following relationship to time:

$$v = \frac{ds}{dt} = 3Tt - 3t^2$$

The curve of this function is shown in Fig. 6.3. We see that, at times 0 and T, v is zero, in other words the marble is at rest at those times. Also, it is clear that the velocity is never constant.

The velocity–time function can, of course, be differentiated, and we have

$$\frac{dv}{dt} = 3T - 6t \tag{6.5}$$

but since the velocity–time function was obtained from the distance–time function by differentiation, it is also correct to write

$$\frac{d^2 s}{dt^2} = 3T - 6t \tag{6.6}$$

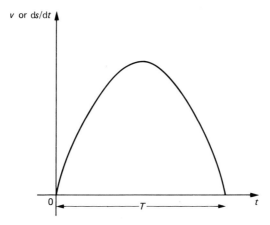

Fig. 6.3   The velocity-time curve, corresponding to the curves in Fig. 6.1(a) and (b), whose equation is $v = 3t(T - t)$.

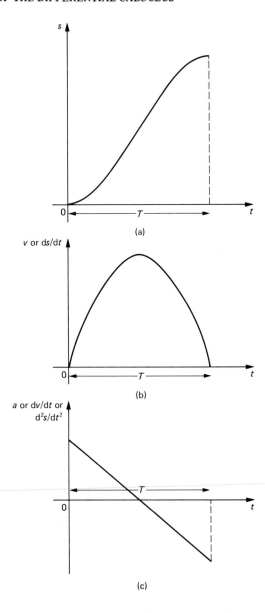

Fig. 6.4   The distance–time, velocity–time, and acceleration–time curves of Figs. 6.1 and 6.3 shown with the same time scale.

In other words, the function of t, $3T - 6t$, can be regarded as either the first-order differential coefficient of the velocity–time function, or the second-order differential coefficient of the distance–time function describing the motion of the marble. Since dv/dt is a first-order differential coefficient (of velocity as a function of time), it must denote the rate of change of velocity, i.e. acceleration. If we examine 6.5 we see that it represents a straight line with intercept $3T$ and a slope of -6; it is, in fact, the acceleration–time relationship that we postulated at the beginning of this discussion (Fig. 6.1(a)).

Bearing in mind that a velocity is a rate of change of one variable with respect to another, and that acceleration is rate of change of velocity, this means that, in general terms, a second order differential coefficient must be the **rate of change of the rate of change** of a function with respect to its variable. In symbols, if $y = f(x)$, then $dy/dx = f'(x)$ is the rate of change of y with respect to x, and $d^2y/dx^2 = f''(x)$ is the rate of change of the rate of change of y with respect to x. The latter is also called the **acceleration** of y with respect to x; if d^2y/dx^2 is a negative quantity, it is often called retardation. To facilitate comparisons, the distance–time, velocity–time and acceleration–time graphs are all shown on a common time-axis in Fig. 6.4.

THE GROWTH OF A MICROBIAL CULTURE

Cell number as a function of time

We shall now begin discussing a biological phenomenon to which some of the concepts of the differential calculus can be applied.

Suppose we inoculated a known volume of a suitable nutrient medium with a species of bacterium. After mixing to ensure a homogeneous dispersion of the bacterial cells, a known small volume of the suspension is sampled and the number of cells contained therein is estimated. By knowing the volume of the sample and the volume of the nutrient medium in which the inoculum was dispersed, it is easy to obtain an estimate of the total number of cells in the inoculum; give this number the symbol n_0. Now incubate the culture and periodically sample it, as described above, so as to obtain numbers of bacterial cells, n, at a series of times, t. These observations can be plotted on a graph, where $t = 0$ is the time at which the first sample is taken. This graph will normally take a form similar to that shown in Fig. 6.5(a). The fundamental question then arises, 'why does the growth of the culture take this particular form?'

Now it is reasonable to suppose that if we had taken many more samples at closer intervals of time, the extra points on the graph would have been situated between the points that are present on Fig. 6.5(a) with respect to both times and numbers. Taking a small portion of the total volume is called **sampling**, and the small volume actually taken is a **sample**.

Consider a particular sampling time. The actual sample taken will contain a number of bacterial cells, indicated by the position of the point on the graph. But if at this particular time a different portion of the culture (different sample) had been taken, it is very unlikely that the number of bacteria counted would have been identical with the number in the actual

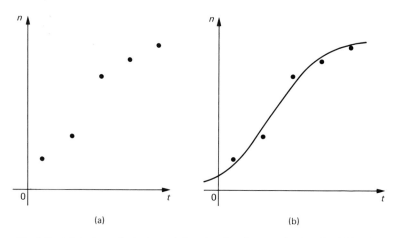

(a)   (b)

Fig. 6.5   (a) An hypothetical set of data showing the growth of a microbial culture.
(b) The same, but with a fitted function.

sample. At this particular time a whole series of samples could be taken, and each would indicate a slightly different value of n. Variation of this kind is evidently bound up with the process of sampling, and has nothing to do with the complete entity (bacterial culture in this case); so the name **sampling variability**, or sampling error, is given to this phenomenon. A common practice, however, is to take only a single sample at any one time, and the assumption is made that, aside from sampling error, the points on the graph would lie on a smooth curve, as shown in Fig. 6.5(b), which we can infer has an underlying functional relationship of the form $n = f(t)$. The question which now arises is, 'what is the form of $f(t)$?'

A growth model based on rates of change—differential equations

There are a number of functions which, from an empirical point of view, would provide good descriptions of the increase of cell number with time in our culture solution; but arbitrarily selecting one of these equations would not help us to elucidate the underlying mechanisms of growth of the bacterial culture. Instead, we attempt to construct a simple mathematical model of growth, by marshalling together relevant facts that we think are already established and putting them into quantitative form. For many biological situations, the present one included, it is natural to formulate the model in terms of rates of change.

This kind of approach, involving the construction of a mathematical model, was referred to in Chapter 1. It was stated that the results produced by the model may or may not agree with experimental findings; if they do not it means that one or more of the assumptions made in constructing the model are untenable. In the following model of the growth of a bacterial culture, although the assumptions made appear to be plausible, the final result of the model, a function of the form $n = f(t)$, does not quite agree with growth curves obtained by careful experimental work. This, however, is not important for our present purpose, which is to show how a mathematical model of a biological process may be constructed using the concepts of the differential calculus. The model is broadly correct and, as will be seen, over-simplification in one of the assumptions is the cause of the discrepancy.

Considering the multiplication of bacterial cells, it is known that in a constant but unlimited environment each cell, on average, divides at a constant rate. Hence the more cells there are present, the more rapid will be the rate of increase for the entire colony. Hence, we can write

$$\frac{dn}{dt} \propto n \qquad (6.7)$$

i.e. the rate of increase of cell number at an instant of time is proportional to the number of cells present at that time. The proportionality sign can be changed for an 'equals' if a constant of proportionality is introduced on the right-hand side of 6.7:

$$\frac{dn}{dt} = kn \qquad (6.8)$$

Notice that: (i) 6.8 takes the form $dn/dt = f(n)$ rather than $dn/dt = f(t)$: the rate of change of cell number is a function of cell number, not time; (ii) in (i) we have written $dn/dt = f(n)$ rather than $dn/dt = f'(n)$, since the

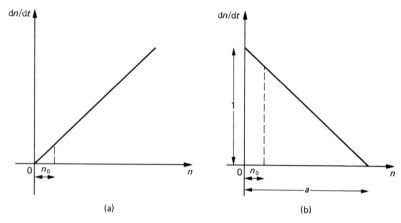

Fig. 6.6   (a) The straight line $dn/dt = kn$. (b) The straight line $dn/dt = 1 - n/a$.

latter would imply that we had started off with $f(n)$ and differentiated it to $f'(n)$.

Equation 6.8 represents a straight line on a graph whose horizontal axis is n, and whose vertical axis is dn/dt. The line passes through the origin (intercept of zero), has a gradient of k, and is shown in Fig. 6.6(a). Evidently, the higher the value of k, the greater the rate of growth of the culture for a given number of cells, n. Notice also that the relationship is valid at the origin, which means that when no organisms are present, the rate of increase is zero — a very sensible conclusion! In fact, the original inoculum would contain a relatively small number of cells, n_0, which, substituted into 6.8, gives a relatively small growth rate for the culture. The gradient, k, is specific to a given species of bacterium under defined environmental conditions.

Now, of course, there is no such thing as an unlimited environment, because this would mean that one bacterial cell would have infinite space and resources at its disposal. For this example, however, we can say that at the time of inoculation the environment is *relatively* unlimited, and as the organisms multiply they use up the nutrients and secrete waste products, some of the latter being actually toxic to the cells. Hence, we can postulate that as the culture grows, the rate of increase is declining; moreover, it is evident that the decline in growth rate must be some function of the (increasing) number of cells, i.e.

$$\frac{dn}{dt} = f(n) \tag{6.9}$$

What form can $f(n)$ take to comply with our theoretical requirements? The simplest form of $f(n)$ that is suitable is shown graphically in Fig. 6.6(b) — a straight line with a negative slope.

Equation 6.8 is a quantitative statement of an intrinsic property of the microbial culture, while equation 6.9 represents the sum total of increasing environmental restrictions. The actual growth rate of the culture at any instant will be the resultant of both processes represented by 6.8 and 6.9, so we can write

$$\frac{dn}{dt} = kn \cdot f(n) \tag{6.10}$$

that is, the growth rate of the culture at any instant is the product of the two processes. We have said that $f(n)$ is a linear function, but we have not yet specified values for its slope and intercept. When n is very small (say 1), the environment is virtually unlimited, and if n could be zero, then the environment would be truly unlimited. Hence, when n is very small, dn/dt is essentially given by 6.8, and this means that $f(n)$ at this time must be approximately unity. So the intercept of the straight line can be given the value of 1. Quantification of the gradient is rather more arbitrary. Let us specify the degree of tolerance to declining nutrients and accumulating wastes as a; then if a is large (i.e. there is a high tolerance) the decline of dn/dt with increasing n is slow, and this is associated with a small negative slope of the straight line. Conversely, if the organisms show only a low tolerance to an increasingly adverse environment, a is small, and the gradient of the line is steep. Evidently, the gradient can be specified as $-1/a$, and so 6.9 becomes

$$\frac{dn}{dt} = 1 - \frac{1}{a}n \tag{6.11}$$

and 6.10 becomes

$$\frac{dn}{dt} = kn \left(1 - \frac{n}{a} \right) \tag{6.12}$$

Looking first at 6.11, if we put $n = a$, $dn/dt = 0$. Hence, the line intersects the horizontal axis at a distance of a from the origin. At the point of intersection dn/dt is zero, and the growth rate has been slowing down as $n \to a$. This means that $n = a$ is the greatest number of cells that can occur in this culture under these defined conditions, and so the constant a has a second meaning in relation to the growth of a bacterial culture.

Turning our attention now to equation 6.12, which quantifies the overall relationship between rate of growth and culture density, multiply out

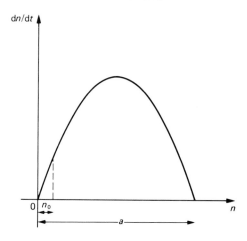

Fig. 6.7   Curve of the second degree polynomial $\mathrm{d}n/\mathrm{d}t = kn(1 - n/a)$.

the bracket:

$$\frac{\mathrm{d}n}{\mathrm{d}t} = kn - \frac{k}{a}n^2 \qquad (6.13)$$

In this form, it is evident that $\mathrm{d}n/\mathrm{d}t = f(n)$ is a second degree polynomial. Referring to the general equation of a second degree polynomial (page 55), we see that the constant term is missing in 6.13, and hence the curve passes through the origin; also, when $n = a$ in 6.12 or 6.13, $\mathrm{d}n/\mathrm{d}t$ is zero. The coefficient of n^2 is negative, hence the curve lies as shown in Fig. 6.7 (c.f. Fig. 4.1(b)). The rate of growth of the culture rises at first while the environment is not very restrictive; loosely, we can say that 6.8 is mainly controlling growth rate during this phase, and the resistance of 6.11 is small. However, the effect of the latter is increasing, while the effect of the former stays constant. So the overall growth rate rises to a maximum and then declines, as function 6.11 approaches zero and thus exerts an ever-increasing effect over the overall growth rate, as described by 6.12.

Equations such as 6.8, 6.11, and 6.12, are known as **differential equations,** because they contain a differential coefficient; so in general **a differential equation describes a rate of change.** The differential equations of this example are of the form $\mathrm{d}n/\mathrm{d}t = \phi(n)$; that is, the instantaneous rate of growth of the culture is a function of culture density at that instant. We commenced this discussion by looking for a function of the form $n = f(t)$, which would tell us the number of cells in the culture at any given time. Now it is possible to obtain $n = f(t)$ from $dn/dt = \phi(n)$, and such a procedure is referred to as 'solving the differential equation'.

Differential equations and their solution form a very important, and difficult, part of mathematics. We are not yet in a position to solve even the simplest equations that have arisen in the preceding discussion. The core procedure in the solution of a differential equation is the reverse process of differentiation, namely, integration; but even when we have considered integration and its methods in the next chapter, there are other topics to be dealt with before we are able to solve an equation like 6.12, and we shall return to the topic later in the book (Chapter 13).

Two final points must be made about the model of bacterial culture growth that has been described. In Fig. 6.5(b), the culture has been assumed to grow according to a smooth curve, implying that cell number can take fractional values, which is clearly wrong. In fact, the actual form of curve is a series of steps, each step corresponding to a cell division; but the curve shown in Fig. 6.5(b) would go through the series of steps, and is merely an approximation (Fig. 6.8). If the number of organisms was relatively few, such as in a population of higher plants or animals, the curve would be only a gross approximation; but in the case of the microbial colony, which involves thousands or millions of individuals, the approximation is good.

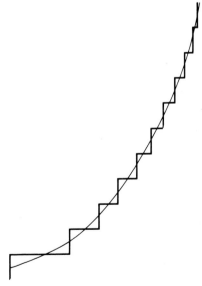

Fig. 6.8   A smooth curve approximating a 'step function'.

Secondly, it was indicated at the beginning of this discussion (page 105) that the result of this model, which is a function of the form $n = f(t)$, does not accord with experimental observation. This means that the curve of the function $n = f(t)$ is somewhat dissimilar to a curve produced from an actual growth experiment in which sampling is frequent. Although the curve produced from an experiment is broadly sigmoid in form (as is the curve of the function $n = f(t)$ derived from the differential equation 6.12 close inspection shows it to have three distinct phases: a lag phase at the beginning where very little increase in cell number occurs, a relatively prolonged phase of growth in which the rate of increase of cell number is proportional to the number present (as described by equation 6.8), and only a short phase of growth rate decline. Evidently, our first assumption embodied in equation 6.8 is satisfactory; but our second is open to question. Broadly, equation 6.9 is correct, but to interpret $f(n)$ as a simple linear function as shown in equation 6.11 is wrong, and a search must be made for another, more realistic and more complex, form of $f(n)$ which gives a result for the model that accords better with experimental findings (see Exercise 1 at the end of this chapter).

MAXIMA AND MINIMA

Maximum growth rate of the microbial culture

In the previous discussion of the growth of a bacterial culture we saw that the growth rate increased while the number of individuals was small, but decreased when the number of organisms was relatively high. This means that the rate of increase must rise to some maximum value before the decline in growth rate sets in (Fig. 6.7). Two questions may be asked: (i) what is the maximum growth rate, and (ii) how dense is the culture when its growth rate is at the maximum? Assuming we knew the values of the constants, a and k, approximate answers to these questions could be obtained by drawing an accurate graph. But it is much quicker, and more accurate, to utilize the differential calculus for this purpose. We shall do this for our microbial culture in this section, and then proceed to a more general discussion of maximum and minimum points.

Referring to Fig. 6.7, it can be seen that the gradient of the curve, that is the rate of change of the growth rate (acceleration), is positive when n is small, and negative when n is large after the maximum growth rate. As n approaches the point where the maximum growth rate occurs, the gradient gets smaller and smaller. At the point of maximum growth rate the gradient of the curve is zero, and thereafter is increasingly negative. The essential point of the last two sentences is that the gradient of the curve is zero when dn/dt is at its maximum, and since in general terms the curve represents a function of the form

$$\frac{dn}{dt} = f(n),$$

which differentiated with respect to n gives

$$\frac{d(dn/dt)}{dt} = f'(n),$$

this implies that

$$f'(n) = 0$$

at the maximum point. Differentiating 6.13 with respect to n:

$$\frac{d(dn/dt)}{dn} = k - \frac{2k}{a} n \qquad (6.14)$$

and on equating the right-hand side of 6.14 to zero, we have

$$k - \frac{2k}{a} n = 0$$

i.e. $k(1 - 2n/a) = 0$, which yields the result $n = a/2$. Hence the maximum growth rate occurs when the colony is half-grown. To find what the maximum rate of growth actually is, we substitute $a/2$ for n in 6.13. This gives

$$\frac{dn}{dt} = k\left(\frac{a}{2}\right) - \frac{k}{a}\left(\frac{a}{2}\right)^2$$

$$= \frac{ka}{2} - \frac{ka^2}{4a} = \frac{ka}{2} - \frac{ka}{4}$$

i.e.

$$\frac{dn}{dt} = \frac{ka}{4}$$

The curve of growth rate against cell number is reproduced in Fig. 6.9, showing the quantities just calculated.

Maximum and minimum, general

Consider the curve in Fig. 6.10, which depicts a function $y = f(x)$ that has several maximum and minimum points. It is important to note exactly what constitutes a maximum or a minimum point (on a curve). A **maximum**

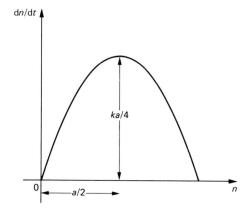

Fig. 6.9   Curve of the second degree polynomial $dn/dt = kn(1 - n/a)$, showing the maximum value of dn/dt, and the value of n at which the maximum occurs.

point is one which has a greater y-value than any *adjacent points* on the curve; hence A and C are *both* maximum points. A **minimum point** is one which has a lower y-value than any adjacent points on the curve; thus B and D are both minimum points. Evidently, maximum and minimum points are 'local phenomena', and each is defined for its own part of the curve. There is no question of a maximum point being necessarily the highest y-value the curve can attain anywhere, although in many instances this may happen to be true; and conversely for a minimum point. In Fig. 6.10, the minimum point D has a greater y-value than the maximum

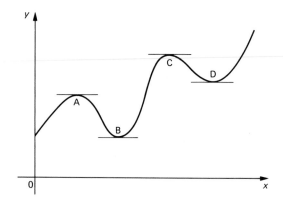

Fig. 6.10   Maxima and minima on a curve.

point A; but in Fig. 6.9, in which the curve is a second degree polynomial, there is no minimum point and the maximum point does indeed represent the highest value that the dependent variable can attain.

Maxima and minima are collectively known as **turning points**, because the gradient of the curve changes from positive to negative or *vice versa* at these locations.

EXAMPLE 6.1

Find the turning points on the curve of the function

$$y = x^3 - 2x^2 - 4x + 1$$

The curve of the function

$$y = x^3 - 2x^2 - 4x + 1 \qquad (6.15)$$

is a cubic polynomial, and evidently there is one maximum and one minimum point (Fig. 6.11(a)). To find the positions of these points, we differentiate the function and equate the differential coefficient to zero

$$\frac{dy}{dx} = 3x^2 - 4x - 4$$

hence

$$3x^2 - 4x - 4 = 0 \qquad (6.16)$$

The relationship 6.16 is a quadratic equation which can be solved by the formula 4.2 applied to the quadratic equation 4.1. In the present instance, we have

$$x = \frac{4 \pm \sqrt{\{(-4)^2 + 48\}}}{6}$$

$$= \frac{4 \pm \sqrt{64}}{6} = \frac{4 \pm 8}{6}$$

Hence

$$x = 2 \quad \text{or} \quad x = -\frac{2}{3}$$

Substituting these x-values successively back into the original equation 6.15,

$$y = (2)^3 - 2(2)^2 - 4(2) + 1$$
$$= 8 - 8 - 8 + 1 = -7$$

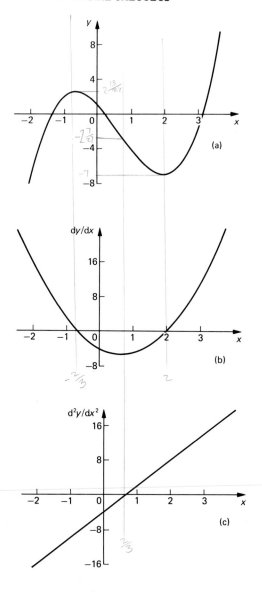

Fig. 6.11   (a) Curve of the function $y = x^3 - 2x^2 - 4x + 1$.
(b) Curve of the function $dy/dx = 3x^2 - 4x - 4$.
(c) The straight line $d^2y/dx^2 = 6x - 4$.

and

$$y = \left(-\frac{2}{3}\right)^3 - 2\left(-\frac{2}{3}\right)^2 - 4\left(-\frac{2}{3}\right) + 1$$

$$= -\frac{8}{27} - \frac{8}{9} + \frac{8}{3} + 1$$

$$= \frac{-8 - 24 + 72 + 27}{27} = \frac{67}{27}$$

$$= 2\frac{13}{27}$$

So the points $(2, -7)$ and $(-\frac{2}{3}, 2^{13}/_{27})$ are turning points of the curve of $y = x^3 - 2x^2 - 4x + 1$. *points plotted on Fig 6.11a*

To determine which of the turning points is a maximum and which is a minimum, without looking at the graph, it is necessary to consider the second derivative. On viewing Fig. 6.11(a) again, we see that around a maximum point the gradient is at first positive but becomes less so as the maximum point is approached; is zero at the point, and then becomes increasingly negative when the maximum point is passed. In other words, the gradient is continually declining in absolute value through a maximum point, which means that the rate of change of the gradient $(d^2 y/dx^2)$ is negative. The converse applies at a minimum point where $d^2 y/dx^2$ is positive. The determination of whether a particular turning point is a maximum or a minimum, then, resolves itself into examining the value of the second-order differential coefficient at that point.

EXAMPLE 6.2

Determine the nature of the turning points of the function in Example 6.1.
We have

$$y = x^3 - 2x^2 - 4x + 1$$

and

$$\frac{dy}{dx} = 3x^2 - 4x - 4$$

hence

$$\frac{d^2 y}{dx^2} = 6x - 4 \tag{6.17}$$

The two turning points are at $x = 2$ and $x = -2/3$, so substituting successively for x into 6.17, we have

$$\text{for } x = 2 \qquad \frac{d^2 y}{dx^2} = 6(2) - 4 = 8$$

$$\text{for } x = -\frac{2}{3} \quad \frac{d^2 y}{dx^2} = 6\left(-\frac{2}{3}\right) - 4 = -8$$

When $x = 2$, $d^2 y/dx^2$ is a positive quantity, and so the point $(2, -7)$ is a minimum. When $x = -2/3$, $d^2 y/dx^2$ is a negative quantity, and so the point $(-2/3, 2^{13}/27)$ is a maximum. These statements can be checked by reference to Fig. 6.11(a).

The relationships between the functions $y = f(x)$, $dy/dx = f'(x)$, and $d^2 y/dx^2 = f''(x)$ for the above example are illustrated in Fig. 6.11. In Fig. 6.11(a) is shown the curve of the original function $y = f(x)$, and in Fig. 6.11(b) is shown the curve of the function giving the differential coefficient of $f(x)$ on the same scale of x. Thus for any value of x, the ordinate in Fig. 6.11(b) is the actual value of the gradient of the curve in Fig. 6.11(a) at that x-value. As already remarked, the gradient of $y = f(x)$ at a turning point is zero, and in this example the turning points are at $x = 2$ and $x = -2/3$. Thus we see in Fig. 6.11(b) that the curve of dy/dx intersects the horizontal axis at these two values of x. Working from left to right along the x-axis, before the maximum point the gradient of $y = f(x)$ is positive, and so the curve of dy/dx is above the x-axis. Similarly, after the minimum point on $y = f(x)$ the gradient is likewise positive, and so the curve of dy/dx is above the x-axis. Between the maximum and minimum points, however, the gradient of $y = f(x)$ is negative, and so over this range of x, i.e. $-2/3 < x < 2$, the curve representing dy/dx is below the x-axis.

Points of inflexion

Look again at Fig. 6.11(a) and recollect how the gradient of a curve changes around a maximum and around a minimum point. We have seen that around a maximum point the rate of change of gradient is negative (Fig. 6.11(b)); after the maximum point is passed, the gradient itself is negative and becomes more so. Before the minimum point, however, we have seen that although the gradient is still negative, it becomes less so as the minimum point is approached (Fig. 6.11(b)). Evidently there must be some point between the maximum and the minimum where the gradient stops becoming more and more negative, and commences to be less negative as one moves along the x-axis from left to right. Such a point is known as a point of inflexion and, unlike maxima and minima, it is not possible to obtain a reasonable approximation of the position of a point of inflexion by graphical methods; the calculus must be used.

We have just said that at a point of inflexion the gradient of the curve ceases becoming more negative and starts to become less negative (i.e. more positive). While the gradient is becoming more negative, the rate of change of gradient, d^2y/dx^2, is negative. When the gradient is becoming less negative, i.e. more positive, d^2y/dx^2 is positive. So at the point of inflexion, d^2y/dx^2 changes from being a negative quantity to a positive one, i.e. $d^2y/dx^2 = 0$ (Fig. 6.11(c)). This discussion is based on the example illustrated in Fig. 6.11, where the maximum of the curve precedes the minimum. If a minimum preceded a maximum, the above comments describing the rate of change of the gradient would be reversed, but nevertheless, the key property of a zero second derivative at the point of inflexion would remain.

EXAMPLE 6.3

Determine the co-ordinates of the point of inflexion of the function employed in Examples 6.1 and 6.2. We have, 6.17,

$$\frac{d^2y}{dx^2} = 6x - 4$$

$6x - 4 = 0$
$6x = 4$
$x = 4$
$x = \frac{2}{3}$

Equating this to zero, gives $x = 2/3$. Substituting back in the original function,

$$y = \frac{8}{27} - \frac{8}{9} - \frac{8}{3} + 1 = -2\frac{7}{27}$$

Hence, the point of inflexion is $\left(\frac{2}{3}, -2\frac{7}{27}\right)$ Plotted on 6.11a

The relationship between d^2y/dx^2, dy/dx, and $y\{= f(x)\}$ can be appreciated by comparing Fig. 6.11(c) with (b) and (a).

At a first reading, the remainder of this chapter can be omitted.

AN ITERATIVE METHOD FOR ESTIMATING THE ROOTS OF A FUNCTION

In Chapter 4, the roots of a function of x were defined as the value of x when $f(x) = 0$. Thus, solving a quadratic equation $a + bx + cx^2 = 0$, involves finding the roots of $f(x) = a + bx + cx^2$ or $y = a + bx + cx^2$. Determining the roots of a quadratic function is quite straightforward (page 56); finding the roots of a cubic or quartic equation is a very much more laborious procedure, and for polynomials of higher degree it is not possible to obtain their roots directly. Neither can the roots of many other types of mathematical function be found directly. In most instances, however, methods are available for obtaining roots of functions approximately, and one such — the Newton–Raphson method — is described here, since it involves differentiation of the function concerned. The term 'iterative' will be explained later on.

Consider the situation shown in Fig. 6.12, in which is shown a graph of the function $y = f(x)$. The curve intersects the x-axis at the point P, distant X from the origin, and this represents a root of the function (there may or may not be others); so that when $x = X$, then $f(x) = 0$; i.e. $f(X) = 0$. The problem is to find the value of X.

Assume that a guess of the value can be made, X_0, represented on the diagram as point Q. Erect a vertical line from Q to intersect the curve of $f(x)$ at R (the length of QR is Y_0, since R lies on the curve of $y = f(x)$); and finally, draw in the tangent to the curve at R, extending it to intersect the x-axis at S. Let point S be distant X_1 from the origin. Now in triangle QRS, by elementary trigonometry

$$\tan \theta = \frac{QR}{QS} = \frac{Y_0}{X_0 - X_1} \qquad (6.18)$$

Now $Y_0 = f(X_0)$, and $\tan \theta = f'(X_0)$. The latter is because $\tan \theta = f'(x)$ at the point where $x = X_0$. Hence, from the extreme left- and right-hand sides of 6.18, we have

$$f'(X_0) = \frac{f(X_0)}{X_0 - X_1}$$

and re-arrangement gives

$$X_1 = X_0 - \frac{f(X_0)}{f'(X_0)} \qquad (6.19)$$

In Fig. 6.12 it is evident that point S, where $x = X_1$, is nearer to point P, where $x = X$ (the actual root being sought), than is the initial guess of the root, $x = X_0$ represented by point Q. Relationship 6.19 shows that X_1, which is a better estimate of the actual root ($x = X$) than is the initial guess ($x = X_0$), can be obtained from X_0 by dividing the value of the function at X_0 by the value of the first differential coefficient at X_0 and subtracting the quotient so obtained from X_0. In general terms, X_0 is called a **starting value**, X_1 the **adjusted value**, and the process of obtaining

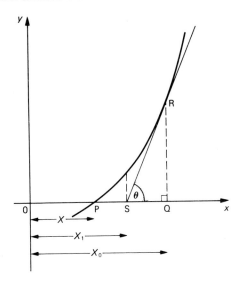

Fig. 6.12   The Newton—Raphson method for approximating the root of a function
(see page 118).

X_1 from X_0 an **iteration**. A starting value has to be provided before the
Newton—Raphson process can be used. Occasionally, a starting value can
be provided by some other mathematical method, but a more generally
useful procedure is to sketch a graph of the function. The iterative process,
that is the procedure summarized in equation 6.19, may be repeated as
often as desired using the adjusted value of the previous iteration as the
starting value of each new iteration. It will usually be observed that the
difference between starting and adjusted values decreases with successive
iterations, so that although the required root cannot be obtained *exactly*
by this procedure, it can be approximated to any desired degree of
accuracy. When the root of a function can be found exactly (e.g. by
applying the formula 4.2 to find the roots of a quadratic function 4.1) the
function is said to have an **explicit solution**; otherwise an iterative process,
such as the Newton—Raphson technique, has to be employed to obtain an
approximate solution.

EXAMPLE 6.4
    Find the roots of the function $f(x) = x^3 - 2x^2 - 4x + 1$. This is the
function of Examples 6.1 to 6.3 above, where our concern has been with
turning points rather than with roots. Before the Newton—Raphson tech-

nique can be used, starting values must be provided, and we shall sketch the curve of the function to assist in this task.

The phrase 'sketch the curve', as opposed to the expression 'draw a graph of the curve', implies that we only require a rough sketch of the curve to show its main features. First, some points on the curve must be calculated, and this is conveniently done as shown in the Table below.

x	x^2	x^3	$-2x^2$	$-4x$	$f(x) = x^3 - 2x^2 - 4x + 1$
3	9	27	−18	−12	−2
4	16	64	−32	−16	17
−2	4	−8	−8	8	−7

In considering values on the curve to be plotted, first remember that we have three points already from Example 6.1 and 6.2: minimum $(2, -7)$, maximum $(-\frac{2}{3}. 2^{13}/_{27})$, inflexion $(\frac{2}{3}, -2\frac{7}{27})$. These are plotted on to the sketch first. Also, inspection of the equation shows that the intercept is 1, i.e. the point $(0, 1)$ lies on the curve. Through these four points, sketch the curve from the maximum to the minimum. Evidently, there is one root near $x = 0$.

Now consider the right-hand side of the graph. Calculation of points on the curve at $x = 3$ and $x = 4$, followed by sketching the curve in this region, shows there to be a root near $x = 3$; and finally, on the left-hand side of the graph, we find a root near $x = -1$. The completed sketch is shown in Fig. 6.13.

In employing equation 6.19 to estimate the roots, note that

$$f(x) = x^3 - 2x^2 - 4x + 1$$

and

$$f'(x) = 3x^2 - 4x - 4$$

The use of the Newton–Raphson technique will be demonstrated in finding the middle root of the equation, near $x = 0$, and this figure will be used as the starting value. So, we have $X_0 = 0$, $f(0) = 1$, and $f'(0) = -4$. Substituting in 6.19:

$$X_1 = 0 - \frac{1}{-4} = \frac{1}{4} = 0.25$$

This is one iteration, in which the initial 'guess' of the root, $x = 0$, has been 'improved' to $x = 0.25$. Now use this value of x as the starting value for a second iteration: $X_1 = 0.25$, $f(0.25) = 0.015625 - 0.1250 - 1 + 1 =$

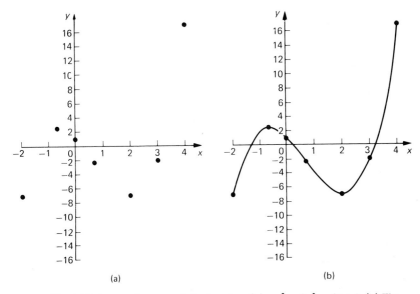

Fig. 6.13   (a) Points on the curve of the function $f(x) = x^3 - 2x^2 - 4x + 1$. (b) The same, but with curve drawn through the points.

-0.109375, and $f'(0.25) = 0.1875 - 1 - 4 = -4.8125$. Substituting in (6.19), we have

$$X_2 = 0.25 - \frac{-0.109375}{-4.8125} = 0.25 - 0.0227 = 0.227273.$$

For a further iteration, we have $X_2 = 0.227273$, $f(0.227273) = 0.011739 - 0.103306 - 0.909092 + 1 = -0.000659$ and $f'(0.227273) = 0.154959 - 0.909092 - 4 = -4.754133$. So

$$X_3 = 0.227273 - \frac{-0.000659}{-4.754133} = 0.227273 - 0.000139 = 0.227134$$

You will notice that the first three places of decimals have remained unchanged between X_2 and X_3; thus, one of the roots of the function $f(x) = x^3 - 2x^2 - 4x + 1$ is 0.227 (correct to 3 decimal places).

EXERCISES 6

1. At the end of the section of this chapter describing the growth of a microbial culture, it was pointed out that while the assumption embodied

in equation 6.8 was satisfactory, the assumption quantified in equation 6.11 was untenable. Consider the following three model equations:

$$\frac{dn}{dt} = 0.2n \tag{6.20}$$

$$\frac{dn}{dt} = 1 - \frac{n}{1\ 000\ 000} \tag{6.21}$$

$$\frac{dn}{dt} = 1 - \left(\frac{n}{1\ 000\ 000}\right)^2 \tag{6.22}$$

and draw graphs of these relationships in the range $0 \leqslant n \leqslant 1\ 000\ 000$. Combine equations 6.20 and 6.21 to give the complete growth model (c.f. equations 6.12 and 6.13). Calculate the co-ordinates of the maximum point, and draw the graph of the complete growth model.

Similarly, combine equations 6.20 and 6.22 to form a complete growth model; calculate the co-ordinates of the maximum point, and draw the graph on the same axes as the previous model. What are your conclusions?

2. Find the co-ordinates of the maximum and minimum points, and of the point of inflexion of the function $f(x) = x^3 - 3x + 1$.

3*. Determine the number of real roots of the cubic polynomial $x^3 - 3x + 3$ (see Chapter 4, page 60), and use the Newton–Raphson technique to determine the root(s) correct to 4 decimal places.

7

Integration: as the Converse of Differentiation

In this chapter we begin studying the other branch of calculus, whose core feature is the process of integration. In its basic form, integration is the converse of differentiation, and we examine this aspect of integration first. It has already been mentioned (Chapter 5, page 75) that integration can also be regarded as a summation process, but we defer this equally important aspect of the subject to Chapter 8.

BASIC CONCEPTS

Consider a function of x, say x^3. Then we know that

$$\frac{d(x^3)}{dx} = 3x^2 \tag{7.1}$$

that is, $3x^2$ is the first differential coefficient or first derivative of x^3. Reversing the notation, we say that x^3 is the **integral** of $3x^2$, and we write

$$\int 3x^2 \, dx = x^3 \tag{7.2}$$

In more general terms, we have written

$$\frac{d\{F(x)\}}{dx} = F'(x) \tag{7.3}$$

(page 88). The symbol $F'(x)$ implies that, starting from a known function

$F(x)$, its differential coefficient has been found. Now in situations in which integration is used, we only know the function $F'(x)$, and required to find $F(x)$. Hence, it is better to give $F'(x)$ a different notation, $f(x)$; then (7.3) becomes

$$\frac{\mathrm{d}\{F(x)\}}{\mathrm{d}x} = f(x) \qquad (7.4)$$

Reversing (7.4), c.f. (7.1) and (7.2), we have that

$$\int f(x)\, \mathrm{d}x = F(x) \qquad (7.5)$$

On the left-hand side of relationship 7.5, $\int$ is the **integral sign**, $f(x)$ is known as the **integrand**, and $\mathrm{d}x$ is the **variable of integration**; one can say that $f(x)$ is **integrated with respect to** x. The whole expression $\int f(x)\mathrm{d}x$ is called an **integral**.

A fundamental difference between the processes of differentiation and integration is that the former can be appreciated and carried out from first principles, whereas the latter cannot be done from first principles. Integration is purely the converse of differentiation; while rules exist to enable any function to be differentiated, the same is not true of integration. We often run into serious problems when attempting to integrate very simple-looking functions.

The indefinite integral and the constant of integration

Let $y = x^3$; then $\mathrm{d}y/\mathrm{d}x = 3x^2$ But suppose, instead, that $y = x^3 + 4$; still we find $\mathrm{d}y/\mathrm{d}x = 3x^2$. To be quite general, suppose that $y = x^3 + c$, where c is any constant; again $\mathrm{d}y/\mathrm{d}x = 3x^2$. Since integration is the converse of differentiation, we may therefore write

$$\int 3x^2\, \mathrm{d}x = x^3 \qquad \text{in the first case}$$

or
$$\int 3x^2\, \mathrm{d}x = x^3 + 4 \qquad \text{in the second case}$$

or
$$\int 3x^2\, \mathrm{d}x = x^3 + c \qquad \text{in general;}$$

and c may be *any* constant. This apparent anomaly arises because the differential coefficient of a constant is zero. So if we are given $\mathrm{d}y/\mathrm{d}x = 3x^2$ and are asked to say what y is equal to, we have no way of knowing whether $y = x^3$, or $y = x^3 + 4$, or $y = x^3 + c$ where c is any constant number. To allow for this contingency, we write the result of an integration problem such as

$$y = \int 3x^2 \, dx$$

as

$$y = x^3 + c$$

where c is known as the **constant of integration**. Typically, the value of c is unknown to us, and consequently $\int 3x^2 \, dx$ is an example of an **indefinite integral**. In more general terms, whenever integration is used as the converse of differentiation, one always obtains the indefinite integral; the constant of integration must be allowed for, but it is usually unknown.

Equations 7.4 and 7.5 together serve to define the nature of integration, and can be called 'the principle of integration'. The evaluation of a differential coefficient is a direct operation for which precise rules exist to obtain a unique result. No such rules exist for integration, and in the last resort the function to be integrated must be recognisable as the result of a differentiation of some kind. A check that an integration has been correctly performed is to differentiate the function obtained as a result of the integration, and to ensure that the differentiated function is equal to the original function.

Integration is not, however, entirely the hit-and-miss process implied in the previous paragraph. There are certain results, which are called **standard integrals**, and many methods by which particular functions can be converted to standard integrals, and so integrated. On the other hand, although any function can be differentiated, there are many functions that cannot be integrated; we shall return to this problem later.

INTEGRATION OF SIMPLE FUNCTIONS

The integration of polynomial functions

If $f(x) = x^m / m$, where m is a constant, we know that $f'(x) = mx^{(m-1)}/m = x^{(m-1)}$. In other words, $\int x^{(m-1)} \, dx = x^m / m + c$. Integration has been carried out by adding one to the power of x, that is $(m-1) + 1$, and dividing by the new power, m.

Thus we have the standard integral

$$\int x^n \, dx = \frac{x^{(n+1)}}{n+1} + c \tag{7.6}$$

where n is any real number, except -1. (If n were -1, we should have to divide x^0 by 0, which gives no sensible result.) See 10.24 p 190 for integral to get $n = -1$

EXAMPLE 7.1

Find

(a) $\int x^2\,dx$ (b) $\int dx/x^2$ (c) $\int \sqrt{x} \cdot dx$

(a) We have immediately from 7.6

$$\int x^2\,dx = x^3/3 + c$$

(b)

$$\int dx/x^2 = \int \frac{1}{x^2}\,dx = \int x^{-2}\,dx$$

Hence, on applying 7.6

$$\int x^{-2}\,dx = \frac{x^{(-2+1)}}{-2+1} + c$$

Thus

$$\int dx/x^2 = -1/x + c$$

(c)

$$\int \sqrt{x} \cdot dx = \int x^{1/2}\,dx = \frac{x^{(1/2+1)}}{1/2+1} + c = \frac{x^{3/2}}{3/2} + c$$

Therefore

$$\int \sqrt{x} \cdot dx = 2\sqrt{x^3}/3 + c$$

In differentiation, we found that if we had a function of x which was made up of simpler functions of x added to, or subtracted from, one another, then differentiation of the whole function was carried out by dealing with each term separately; i.e. if

$$f(x) = \phi(x) \pm \psi(x)$$

then

$$f'(x) = \phi'(x) \pm \psi'(x).$$

Since integration is the converse of differentiation, the same rule will apply in integration; hence if

$$f(x) = \phi(x) \pm \psi(x)$$

then

$$\left.\begin{array}{c} \\[1ex] \int f(x)\,dx = \int \phi(x)\,dx \pm \int \psi(x)\,dx + c \end{array}\right\}\qquad (7.7)$$

Also in differentiation, if $f(x) = ax^n$, then $f'(x) = anx^{(n-1)}$, the coefficient a going through the process unchanged; the same is true for a coefficient in integration, and we may therefore place such a constant outside the integral sign:

$$\int ax^n dx = a \int x^n dx = \frac{ax^{(n+1)}}{n+1} + c \qquad (7.8)$$

The centre term in 7.8 consists of a multiplied by the standard integral on the left-hand side of 7.6, so the result of integrating ax^n with respect to x is readily obtained.

EXAMPLE 7.2
Find

(a) $\int 2x^3 dx$ (b) $\int \frac{x\,dx}{2}$ (c) $\int \frac{5\sqrt{x} \cdot dx}{3}$

(a) $\qquad\qquad \int 2x^3 dx = 2\int x^3 dx = 2x^4/4 + c$

Therefore

$$\int 2x^3 dx = x^4/2 + c$$

(b) $\qquad\qquad \int \frac{x\,dx}{2} = \frac{1}{2}\int x\,dx = \frac{1}{2} \times \frac{x^2}{2} + c$

Hence

$$\int x\,dx/2 = x^2/4 + c$$

(c) $\qquad\qquad \int \frac{5\sqrt{x} \cdot dx}{3} = \frac{5}{3}\int x^{1/2} dx = \frac{5}{3} \times \frac{2x^{3/2}}{3} + c$

Thus

$$\int \frac{5\sqrt{x} \cdot dx}{3} = \frac{10}{9}\sqrt{x^3} + c$$

The integration of c^x

We have seen that $d(c^x)/dx = c^x \cdot \log_e c$; therefore

$$\int c^x \cdot \log_e c \cdot dx = c^x + k$$

P84 Eq 5.17

where k is the constant of integration. Since a coefficient may be placed outside an integral sign:

$$\log_e c \int c^x dx = c^x + k \qquad (7.9)$$

Dividing both sides of 7.9 by $\log_e c$, we have

$$\int c^x \, dx = \frac{c^x}{\log_e c} + k' \qquad (7.10)$$

where $k' = k/\log_e c$. Relationship 7.10 is another standard integral (prove it by differentiating the right-hand side), and it is evident that

$$\int ac^x \, dx = a \int c^x \, dx = \frac{ac^x}{\log_e c} + k \qquad (7.11)$$

The integration of $(a + bx)^n$
Another useful result is

$$\int (a + bx)^n \, dx = \frac{(a + bx)^{(n+1)}}{b(n + 1)} + c, \quad n \neq -1 \qquad (7.12)$$

The standard integral 7.12 is similar to that of expression 7.6, and is valid for any value of n except -1. No derivation of the result in 7.12 is given here, because it is simple enough to differentiate the right-hand side and see that this does yield the left-hand side. Indeed, as we have already remarked (page 125), integration may often have to be attempted on such a trial and error basis.

If this is your first acquaintance with the subject of integration, you should now proceed to the beginning of Chapter 8.

METHODS OF INTEGRATION

As already remarked, the process of integration of a function requires first that the function be resolved into a form recognizable as a standard integral. There are several methods, of varying complexity, for attempting this process; hitherto we have only used the trivial ideas of placing a coefficient outside the integral sign, and resolving the integral of a polynomial into a sum of integrals of each individual term.

It must be admitted that integration is usually a difficult process, much more so that differentiation. The methods to be considered in the remainder of this chapter, although simple in principle, are not always easy to apply in practice. There are many functions of biological importance which require these methods for their integration, but it may be preferable to delay consideration of them to a second reading, especially as we shall need to make use of standard integrals presented later in the book.

Partial fractions and their use in integration

If we are given the expression

$$\frac{2}{3-x} + \frac{1}{2+x}$$

it is easy to amalgamate the two fractions, adding them together by putting them over a common denominator:

$$\frac{2(2+x)+(3-x)}{(3-x)(2+x)} = \frac{4+2x+3-x}{(3-x)(2+x)} = \frac{7+x}{(3-x)(2+x)}$$

Thus

$$\frac{2}{3-x} + \frac{1}{2+x} = \frac{7+x}{(3-x)(2+x)} \qquad (7.13)$$

So, working from left to right in 7.13 is easy.

Now the two fractions on the left-hand side can be recognized as standard integrals (equation 10.28 on page 192):

$$\int \frac{2dx}{3-x} = 2 \int \frac{dx}{3-x} = -2 \cdot \log_e (3-x) + c$$

and

$$\int \frac{dx}{2+x} = \log_e (2+x) + c$$

but the right-hand side of 7.13 cannot be recognized as a standard integral as it stands. Hence, if we were required to find

$$\int \frac{7+x}{(3-x)(2+x)} \, dx,$$

it would first of all be necessary to decompose this fraction into its **partial fractions**; in other words, we require a process whereby we may work from right to left in 7.13. Such a method is known simply as 'partial fractions'.

Returning to the above example, we first of all know that the partial fractions of $(7+x)/\{(3-x)(2+x)\}$ will be two in number, and one will have $(3-x)$ in its denominator and the other will have $(2+x)$ in its denominator. The numerator of each partial fraction will be unknown: let us represent these as A in one and as B in the other partial fraction. Then we can write

$$\frac{7+x}{(3-x)(2+x)} \equiv \frac{A}{3-x} + \frac{B}{2+x} \qquad (7.14)$$

The $\equiv$ sign means that relationship 7.14 is an identity, i.e. that the right-hand and left-hand sides must be equal to one another for *all* values of the variable x. (An = sign does not imply this: the equality may hold for only certain values of x.)

Now, multiply both sides of 7.14 by $(3 - x)(2 + x)$, the denominator of the fraction whose partial fractions are required, and we have

$$7 + x \equiv A(2 + x) + B(3 - x) \tag{7.15}$$

Since relationship 7.15 must hold for all values of x, A and B may be determined by selecting appropriate values for x and substituting them successively into 7.15. First put $x = 3$, when it is obvious that the second term on the right-hand side of 7.15 disappears; this leaves only A as an unknown and we may solve for it. Then put $x = -2$, which in a similar fashion leaves only B as an unknown which can be solved for. So, put

$$x = 3: \quad \text{then} \quad 7 + 3 = A(2 + 3)$$
$$\text{i.e.} \quad 10 = 5A \quad \text{and so} \quad A = 2.$$

Put

$$x = -2: \quad \text{then} \quad 7 + (-2) = B\{3 - (-2)\}$$
$$\text{i.e.} \quad 5 = 5B \quad \text{and so} \quad B = 1.$$

Hence the two partial fractions are, from 7.14, $2/(3 - x)$ and $1/(2 + x)$, which we see from 7.13 are correct. Thus, we have that

$$\int \frac{(7 + x)dx}{(3 - x)(2 + x)} = \int \frac{2dx}{3 - x} + \int \frac{dx}{2 + x} = \log_e(2 + x) - 2 \cdot \log_e(3 - x) + c$$

EXAMPLE 7.3
   Integrate

(a) $$\frac{3x^2 - 7}{x(x - 2)(x + 4)}$$

(b) $$\frac{x}{x^2 + 5x + 6}$$

(a) First, the expression is decomposed into partial fractions. Let

$$\frac{3x^2 - 7}{x(x - 2)(x + 4)} \equiv \frac{A}{x} + \frac{B}{x - 2} + \frac{C}{x + 4}$$

(The fact that the numerator of the composite fraction contains a function of x does not affect the method so long as the highest power of x in the numerator is lower than the highest power in the denominator.) Multiply both sides by $x(x - 2)(x + 4)$:

$$3x^2 - 7 \equiv A(x - 2)(x + 4) + Bx(x + 4) + Cx(x - 2)$$

Put

$x = 2$:   then $3(2)^2 - 7 = B(2)(6)$ i.e. $12 - 7 = 12B$

so

$$B = 5/12.$$

Put

$x = -4$:   then $3(-4)^2 - 7 = C(-4)(-6)$ i.e. $48 - 7 = 24C$

so

$$C = 41/24.$$

Put

$x = 0$:   then $-7 = A(-2)(4)$ i.e. $-7 = -8A$

so

$$A = 7/8.$$

Hence

$$\frac{3x^2 - 7}{x(x - 2)(x + 4)} = \frac{7}{8x} + \frac{5}{12(x - 2)} + \frac{41}{24(x + 4)}$$

Now

$$\int \frac{7dx}{8x} = \frac{7}{8} \int \frac{dx}{x} = \frac{7}{8} \cdot \log_e x + c \qquad \text{(page 190)}$$

$$\int \frac{5dx}{12(x - 2)} = \frac{5}{12} \int \frac{dx}{x - 2} = \frac{5}{12} \cdot \log_e(x - 2) + c$$

$$\left. \int \frac{41dx}{24(x + 4)} = \frac{41}{24} \int \frac{dx}{x + 4} = \frac{41}{24} \cdot \log_e(x + 4) + c \right\} \text{(page 192)}$$

Hence we may summarize the whole procedure thus:

$$\int \frac{(3x^2 - 7)dx}{x(x - 2)(x + 4)} = \int \left\{ \frac{7}{8x} + \frac{5}{12(x - 2)} + \frac{41}{24(x + 4)} \right\} dx$$

$$= \int \frac{7dx}{8x} + \int \frac{5dx}{12(x - 2)} + \int \frac{41dx}{24(x + 4)}$$

$$= \frac{7}{8} \int \frac{dx}{x} + \frac{5}{12} \int \frac{dx}{x - 2} + \frac{41}{24} \int \frac{dx}{x + 4}$$

Thus

$$\int \frac{(3x^2 - 7)dx}{x(x - 2)(x + 4)} = \frac{7}{8} \cdot \log_e x + \frac{5}{12} \cdot \log_e(x - 2) + \frac{41}{24} \cdot \log_e(x + 4) + c$$

where c is the sum of the constants obtained for the three individual integrations.

(b) The denominator is a quadratic function of x, and if it can be factorized into the form $(x + a)(x + b)$, where a and b are constants, then the method of partial fractions can be used to integrate the given expression. To factorize $x^2 + 5x + 6$, we require two numbers, a and b, such that $a + b = 5$ and $ab = 6$. The numbers 2 and 3 satisfy these conditions so $x^2 + 5x + 6 = (x + 2)(x + 3)$. Thus the working of the problem in outline is

NB

$$\int \frac{x\,dx}{x^2 + 5x + 6} = \int \frac{x\,dx}{(x + 2)(x + 3)} \quad \text{(by factorization)}$$

$$= \int \left\{ \frac{3}{x + 3} - \frac{2}{x + 2} \right\} dx \quad \text{(by partial fractions)}$$

$$= 3 \int \frac{dx}{x + 3} - 2 \int \frac{dx}{x + 2}$$

Thus

$$\int \frac{x\,dx}{x^2 + 5x + 6} = 3 \cdot \log_e(x + 3) - 2 \cdot \log_e(x + 2) + c.$$

So far, we have considered examples in which the denominator of the fraction to be integrated contains only linear functions of x (i.e. of the form $a + bx$), or functions of x that can be put into linear form, as in Example 5.3(b) above. If this is not the case, the procedure must be very slightly modified.

EXAMPLE 7.4
Find

$$\int \frac{2x\,dx}{(x + 3)^2}$$

Consider first how to split $2x/(x + 3)^2$ into partial fractions. At first sight, it might seem appropriate to let

$$\frac{2x}{(x + 3)^2} \equiv \frac{A}{x + 3} + \frac{B}{x + 3}$$

However, this would be incorrect, since when the next step of multiplying both sides by $(x + 3)^2$ is done, we have

$$2x \equiv A(x + 3) + B(x + 3) \tag{7.16}$$

Now it can be seen that 7.16 is not an identity, true for *all* values of x, since if we put $x = -3$ we have the anomaly that $-6 = 0$. If, however, we let

$$\frac{2x}{(x + 3)^2} \equiv \frac{A}{x + 3} + \frac{B}{(x + 3)^2}$$

then, on multiplying both sides by $(x + 3)^2$, we have

$$2x \equiv A(x + 3) + B \tag{7.17}$$

and 7.17 *is* an identity. Now put $x = -3$; then $2(-3) = B$, and so $B = -6$. Using this value of B in 7.17, we have $2x = A(x + 3) - 6$, i.e. $A(x + 3) = 2x + 6$ or $A(x + 3) = 2(x + 3)$, and so $A = 2$. Thus

$$\frac{2x}{(x + 3)^2} = \frac{2}{x + 3} - \frac{6}{(x + 3)^2}$$

(verify this by working from right to left). Summarizing the whole problem:

$$\int \frac{2x\,dx}{(x + 3)^2} = \int \left\{ \frac{2}{x + 3} - \frac{6}{(x + 3)^2} \right\} dx$$

$$= 2 \int \frac{dx}{x + 3} - 6 \int \frac{dx}{(x + 3)^2}$$

and so

$$\int \frac{2x\,dx}{(x + 3)^2} = 2 \cdot \log_e(x + 3) + \frac{6}{x + 3} + c$$

See page 128 for the integration of the second term.

Integration by substitution

Suppose we were required to find $\int x\sqrt{(1 + x^2)}\,dx$. It is not possible to multiply out the terms first in order to look for a standard integral. However, we try to put part of the integrand equal to another variable in the hope that the resulting integral will be a standard one. In this example, put

$$z = 1 + x^2 \tag{7.18}$$

Now, apart from the integral sign, the integral in this example consists of three terms: x and $\sqrt{(1 + x^2)}$, which together comprise the integrand, and the variable of integration, dx. We cannot merely substitute for one of these components, e.g. $\sqrt{(1 + x^2)}$ $(= \sqrt{z})$, ending up with an integral containing the two variable quantities, x and z; but by substituting for each of the terms in the integral, we must finally obtain an integral entirely in terms of z. Substituting for $\sqrt{(1 + x^2)}$ is easy: it is merely $\sqrt{z}$. Now from 7.18, $x = \pm\sqrt{(z - 1)}$. Differentiating 7.18, $dz/dx = 2x$; i.e. $dx = dz/2x$. But we already have seen that $x = \pm\sqrt{(z - 1)}$. hence $dx = \pm dz/2\sqrt{(z - 1)}$. Thus, we may write

$$\int x \cdot \sqrt{(1 + x^2)} \cdot dx = \int \sqrt{(z - 1)} \cdot \sqrt{z} \cdot \frac{dz}{2\sqrt{(z - 1)}} = \int \frac{\sqrt{z} \cdot dz}{2}$$

$$= \frac{1}{2}\int z^{1/2} dz$$

The last integral is a standard one and gives

$$\frac{1}{2}\left(\frac{z^{3/2}}{3/2}\right) + c = z^{3/2}/3 + c = \sqrt{z^3}/3 + c.$$

Substituting back to x for z, we have that

$$\int x\sqrt{(1 + x^2)}dx = \frac{1}{3}\sqrt{(1 + x^2)^3} + c$$

Integration by substitution is not always a successful technique in · practice; it is often difficult to know exactly what to substitute for what, and after many attempts at different substitutions it may be found that the integral cannot be reduced to a standard form by this method.

EXAMPLE 7.5
   Find

$$\int \frac{(x + 1)dx}{\sqrt{(x^2 + 2x - 1)}}$$

Put $z = (x^2 + 2x - 1)^{1/2}$. By the function of a function rule we have

$$\frac{dz}{dx} = \frac{(x + 1)}{(x^2 + 2x - 1)^{1/2}}$$

i.e.

$$(x + 1)dx = (x^2 + 2x - 1)^{1/2}dz$$

Hence

$$\int \frac{(x+1)dx}{(x^2+2x-1)^{1/2}} = \int \frac{(x^2+2x-1)^{1/2}dz}{(x^2+2x-1)^{1/2}} = \int dz = z + c$$

and so, substituting back for z, we have the result

$$\int \frac{(x+1)dx}{\sqrt{(z^2+2x-1)}} = \sqrt{(x^2+2x-1)} + c$$

Integration by parts

This method is used when it is required to integrate the product of two functions of x, but the procedure can also be used to integrate certain single functions of x for which no other method suffices. The method involves a formula into which the function, or functions, are substituted in a similar manner to the product rule for differentiation. Indeed, we start with the product rule for differentiation to derive the formula for integration by parts.

For two functions of x, $\psi(x)$ and $\theta(x)$, the product rule for differentiation gives

$$\frac{d\{\psi(x) \cdot \theta(x)\}}{dx} = \psi(x) \cdot \frac{d\{\theta(x)\}}{dx} + \theta(x) \cdot \frac{d\{\psi(x)\}}{dx}$$

(c.f. equations 5.24 on page 89).
Re-arranging:

$$\psi(x) \cdot \frac{d\{\theta(x)\}}{dx} = \frac{d\{\psi(x) \cdot \theta(x)\}}{dx} - \theta(x) \cdot \frac{d\{\psi(x)\}}{dx}$$

Integrating with respect to x gives

$$\int \psi(x) \cdot \frac{d\{\theta(x)\}}{dx}dx = \psi(x) \cdot \theta(x) - \int \theta(x) \cdot \frac{d\{\psi(x)\}}{dx}dx + c$$

Put

$$\phi(x) = \frac{d\{\theta(x)\}}{dx}, \quad \text{then} \quad \theta(x) = \int \phi(x)dx$$

The previous line now finally becomes

$$\int \psi(x) \cdot \phi(x)dx = \psi(x) \cdot \int \phi(x)dx - \int \left[\frac{d\{\psi(x)\}}{dx} \cdot \int \phi(x)dx \right] dx + c$$

$$(7.19)$$

Equation 7.19 is the formula for integration by parts. On examining the terms on the right-hand side of 7.19, we see that $\int\phi(x)dx$ appears in each term; consequently, $\phi(x)$ must be integrable by some other method. The function $\psi(x)$ is not required to be integrated: true, it is necessary to differentiate it, but as all functions can be differentiated this is no problem. Hence $\psi(x)$ may be a function which cannot be integrated by other methods.

In using the product rule to differentiate the product of two functions of x, it does not matter which function is identified with $\psi(x)$ or with $\phi(x)$ (see equation 5.24 on page 89); but when integrating by parts, the remarks in the previous paragraph must be borne in mind.

EXAMPLE 7.6

Find

(a) $\int x \cdot \sqrt{(2x+3)^3}\,dx$

(b) $\int (x^2+1) \cdot \sqrt{(2x+3)}\,dx$

(a) Put $\psi(x) = x$ and $\phi(x) = \sqrt{(2x+3)^3}$, then using 7.19

$$\int x \cdot \sqrt{(2x+3)^3}\,dx = x\int\sqrt{(2x+3)^3}\,dx - \int \left\{\frac{d(x)}{dx}\int\sqrt{(2x+3)^3}\,dx\right\}dx$$

(7.20)

Now

$$\int\sqrt{(2x+3)^3}\,dx = \int(2x+3)^{3/2}\,dx$$
$$= \frac{(2x+3)^{5/2}}{2(5/2)} + c$$
$$= \frac{(2x+3)^{5/2}}{5} + c \qquad \text{(see 7.12)}$$

and $d(x)/dx = 1$. Hence 7.20 becomes

$$\int x \cdot \sqrt{(2x+3)^3}\,dx = x(2x+3)^{5/2}/5 - \int(1)(2x+3)^{5/2}\,dx/5 + c$$
$$= x(2x+3)^{5/2}/5 - (2x+3)^{7/2}\ 35 + c$$

Thus

$$\int x \cdot \sqrt{(2x+3)^3}\,dx = \frac{x\sqrt{(2x+3)^5}}{5} - \frac{\sqrt{(2x+3)^7}}{35} + c \qquad (7.21)$$

(b) Put $\psi(x) = x^2 + 1$ and $\phi(x) = \sqrt{(2x + 3)}$, then using 7.19

$$\int (x^2 + 1)\sqrt{(2x + 3)}dx = (x^2 + 1)\int \sqrt{(2x + 3)}dx$$
$$- \int \left\{ \frac{d(x^2 + 1)}{dx} \int \sqrt{(2x + 3)}dx \right\} dx$$

$$(7.22)$$

Now

$$\int \sqrt{(2x + 3)}dx = \int (2x + 3)^{1/2}dx = (2x + 3)^{3/2}/3 + c$$

and $d(x^2 + 1)/dx = 2x$. Hence 7.22 becomes

$$\int (x^2 + 1)\sqrt{(2x + 3)}dx = \frac{(x^2 + 1)(2x + 3)^{3/2}}{3} - \int \frac{2x(2x + 3)^{3/2}}{3} + c$$

i.e.

$$\int (x^2 + 1)\sqrt{(2x + 3)}dx = \frac{(x^2 + 1)(2x + 3)^{3/2}}{3} - \frac{2}{3}\int x(2x + 3)^{3/2} + c$$

$$(7.23)$$

This example is of a type that frequently arises in integration by parts, i.e. the process has to be done twice, or even more. Here, the integral in 7.23 has to be integrated by parts again; however, in this instance it has already been done in Example (a), 7.21, so we have the result

$$\frac{2}{3}\int x(2x + 3)^{3/2}dx = \frac{2x(2x + 3)^{5/2}}{15} - \frac{2(2x + 3)^{7/2}}{105} + c$$

and so, substituting in 7.23, we finally have

$$\int (x^2 + 1) \cdot \sqrt{(2x + 3)}dx = \frac{(x^2 + 1)\sqrt{(2x + 3)^3}}{3}$$
$$- \frac{2x\sqrt{(2x + 3)^5}}{15} + \frac{2\sqrt{(2x + 3)^7}}{105} + c$$

Integration by parts of a single function

When this method is used to integrate a single function, rather than a product of two functions, we begin by taking the function to be integrated as $\psi(x)$ in 7.19 and putting $\phi(x)$ equal to unity. Very useful applications of this method of integrating a single function are given in Example 10.5 on page 193.

EXERCISES 7

1. Find the following indefinite integrals:

(a) $\int 6x^2\,dx$ (b) $\int (5/x^2)\,dx$ (c) $\int (x^3 - a^3)\,dx$ (d) $\int (x - 1/x)^2\,dx$

(e) $\int (1 + x^2) \cdot \sqrt{x}\,dx$ (f) $\int (2x - 1)^5\,dx$ (g) $\int 1/\sqrt[3]{(1 - 4x)}\,dx$

2*. Integrate the following, using the method of partial fractions

(a) $\int \dfrac{(3x + 4)dx}{(x - 2)(x + 3)}$ (b) $\int \dfrac{(x^2 - 3x + 3)dx}{(x - 1)(x - 2)(x - 3)}$

3*. Using the substitutions indicated, find

(a) $\int \dfrac{dx}{x\sqrt{(1 + x^2)}}$ putting $z = \sqrt{(1 + x^2)}$

(b) $\int \dfrac{x\,dx}{(2x + 1)^3}$ putting $z = 2x + 1$

8

Integration: as a Summation

You may be wondering why, if integration is the converse of differentiation, the process is not given a name like 'de-differentiation'. The answer to this lies in the remark made at the beginning of the previous chapter, that the process of integration can be applied in two different situations. We have already examined one of these, the concept of integration as the reverse of differentiation. The other concept is implied in the name of the process, 'integration', which means 'adding together', and we look at this aspect of integration in this chapter.

THE AREA UNDER A CURVE AND THE DEFINITE INTEGRAL

We begin by stating and proving a basic fact. In Fig. 8.1 the area enclosed by the curve of $y = f(x)$, the x-axis, and the vertical lines at $x = a$ and $x = b$ (i.e. the area enclosed by the curve AB and the lines BC, CD, and DA) can be represented by an integral with $f(x)$ as integrand; and moreover, the area may be evaluated precisely.

To prove these statements, look first at Fig. 8.2. It is required to find the area enclosed by the curve of $y = f(x)$, the x-axis, and the vertical lines at $x = a$ and $x = b$. Imagine now that the area is dissected into a very large number of very thin vertical strips, each of width δx. In Fig. 8.2 one of these strips is shown (highly magnified in width) whose left-hand edge PM is distant x from the origin. The height of this edge is y, i.e. $f(x)$, since PM is distant x from the origin. The area of the rectangle PMNR is $y \cdot \delta x$, and, since the area enclosed by PQR is very small, the area of the complete strip PMNQ is approximately equal to $y \cdot \delta x$.

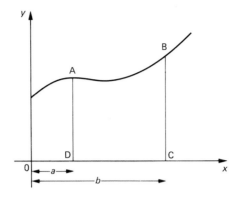

Fig. 8.1   The area under a curve (see page 139).

Now let us define some quantities in Fig. 8.2 and give them symbols. Let the area required, ABCD, be S and let the area APMD be s. Further, since δx is very small, the small element of area given by the strip is, as we have seen, $\eqsim y \cdot \delta x$. Since area PQMN can be regarded as a small increase to area APMD due to a small increase δx in x, we may call the area PQMN δs. Hence

$$\delta s \eqsim y \cdot \delta x \qquad (8.1)$$

The narrower the strips, the smaller δx becomes, and the approximation

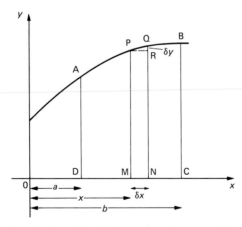

Fig. 8.2   A typical vertical strip in an area under a curve (see page 139).

8.1 becomes more accurate. So we may say that,

$$\text{as} \quad \delta x \to 0, \quad \text{so} \quad \delta s \to y \cdot \delta x$$

or

$$\frac{\delta s}{\delta x} \to y.$$

Writing this in proper limit notation, we have

$$\lim_{\delta x \to 0} \left(\frac{\delta s}{\delta x} \right) = y \tag{8.2}$$

But in keeping with previous notation, equation 5.5, where we wrote

$$\lim_{\delta x \to 0} \left(\frac{\delta y}{\delta x} \right) = \frac{dy}{dx}$$

we can write in the present instance

$$\lim_{\delta x \to 0} \left(\frac{\delta s}{\delta x} \right) = \frac{ds}{dx} \tag{8.3}$$

Since the left-hand sides of 8.2 and 8.3 are equal to one another, we may equate the right-hand sides of these two expressions:

$$\frac{ds}{dx} = y$$

Cross-multiplying gives

$$ds = y \cdot dx$$

and so s, the area APMD, is given by

$$s = \int y dx;$$

or, since $y = f(x)$,

$$s = \int f(x) dx.$$

Now assume that

$$\int f(x) dx = F(x) + c$$

where $d\{F(x)\}/dx = f(x)$ (page 124), and c is a constant of integration; therefore

$$s = F(x) + c \tag{8.4}$$

We may regard the measurement of area s as beginning from the line AD.

Hence at line AD, where $x = a$, there is no area, and so $s = 0$. So substituting for s and x in 8.4, we have

$$0 = F(a) + c$$

which gives

$$c = -F(a)$$

Substituting back into 8.4 for c gives

$$s = F(x) - F(a) \qquad (8.5)$$

Now consider the whole required area ABCD $(=S)$; for this area $x = b$ in 8.5. On substituting for x in 8.5, we have

$$S = F(b) - F(a) \qquad (8.6)$$

Equation 8.6 is the required result, and can be stated as follows: 'the area under the curve of $y = f(x)$, between the limits $x = a$ and $x = b$, is the value of $F(x)$ when $x = b$ minus the value of $F(x)$ when $x = a$; $F(x)$ is the integral of $f(x)$'. Notice that 8.6 contains no constant of integration; S is the difference between $F(b)$, which does contain such a constant, and $F(a)$ which contains the same constant, and thus cancels out.

Evaluation of a definite integral

In working problems on areas under curves, the following notation is generally used. The right-hand side of 8.6 is written as

$$\left[F(x) \right]_a^b \qquad (8.7)$$

that is, the integrated function is put in square brackets and the upper and lower limits of x at which F is to be evaluated are written as shown. The integral of the function of the original curve which gives rise to 8.7 is written as

$$\int_a^b f(x)\mathrm{d}x \qquad (8.8)$$

which is the integral of $f(x)\mathrm{d}x$ as written in the previous chapter, but with the upper and lower limits shown. Therefore, as 8.6, 8.7, and 8.8 are merely different ways of expressing the same thing, we have, in the order they arise in working through a problem

$$S = \int_a^b f(x)\mathrm{d}x = \left[F(x) \right]_a^b = F(b) - F(a) \qquad (8.9)$$

where

$$d\{F(x)\}/dx = f(x).$$

The symbol $\int_a^b f(x)dx$ is clearly somewhat different from the symbol $\int f(x)dx$, for

$$\int f(x)dx = F(x) + c \tag{8.10}$$

and

$$\int_a^b f(x)dx = F(b) - F(a) \tag{8.11}$$

However, in both instances $d\{F(x)\}/dx = f(x)$ and so integration is actually carried out in the same way in each case. But in 8.11 there is the further step of inserting limits into the integrated function. Since 8.11 does not involve the unknown constant of integration, $\int_a^b f(x)dx$ is known as the **definite integral**; $\int f(x)dx$, you will recall, is the indefinite integral.

EXAMPLE 8.1

Find the areas under the curves indicated:

(a) of the function $y = x^2$ between the limits of 1 and 3;
(b) of the function $y = \sqrt{(2x + 1)}$ between limits of 0 and 4.

(a) The situation is shown in Fig. 8.3, and the value of the shaded area is required, S. Since the curve represents the function $y = x^2$, and the vertical limits of the area are $x = 1$, and $x = 3$, the required area is given by

$$S = \int_1^3 x^2 \, dx$$

$$= \left[\frac{x^3}{3} \right]_1^3$$

$$= \frac{(3)^3}{3} - \frac{(1)^3}{3} = 9 - \frac{1}{3} = 8\tfrac{2}{3}$$

Hence the shaded area is $8\tfrac{2}{3}$ square units.

(b) The required area is given by

$$S = \int_0^4 \sqrt{(2x + 1)}dx = \int_0^4 (2x + 1)^{1/2} dx = \left[\frac{(2x + 1)^{3/2}}{3} \right]_0^4 .$$

(See equation 7.12 for the integration of this function.)

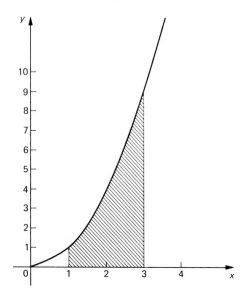

Fig. 8.3   The area under the curve of $y = x^2$ between the limits of $x = 1$ and $x = 3$ (see Example 8.1(a)).

Hence

$$S = \frac{\sqrt{\{2(4) + 1\}^3}}{3} - \frac{\sqrt{\{2(0) + 1\}^3}}{3} = 9 - \frac{1}{3} = 8\tfrac{2}{3} \text{ square units}$$

Another difference should now be evident between a definite and an indefinite integral: while the latter merely gives rise to another function, the former gives a numerical value.

For example,

$$\int x^2 \, dx = x^3/3 + c$$

whereas

$$\int_1^3 x^2 \, dx = 8\tfrac{2}{3}.$$

A definite integral can also exist, or be thought of, on its own without any reference to an area under a curve, just as a differential coefficient does not have to be associated with the gradient of a curve, or a function of x does not have to be identified with a curve in the x–y plane. One can merely say, 'Evaluate $\int_1^3 x^2 \, dx$', and the answer is $8\tfrac{2}{3}$.

It cannot be too strongly emphasized that the foregoing material of this chapter is very important as background theory to many leading aspects of quantitative biology. The discussion, notations involved, and the proof that the area under a curve can be evaluated by a definite integral, should be thoroughly assimilated and understood before proceeding.

'Negative' areas

Consider the areas shown in Fig. 8.4, each of which will be given by $\int_a^b f(x)dx$. In Fig. 8.4(a) and (b), the area is on the upper side of the x-axis (positive y-values); further, in (a) the upper limit, b, is greater than the lower limit, a, i.e. $b > a$. This situation is identical with our previous examples, and as we have seen, the definite integral yields a numerical value which is positive. In (b) is shown the same area, but now the upper limit, b, is less than the lower limit, a, i.e. $b < a$. Under these conditions, the definite integral will yield the same *numerical* result as in (a) but opposite in sign, i.e. negative. Nevertheless, it is obvious that the same area is being measured, as is shown by the result being numerically the same; but since the limits are reversed, the sign of the definite integral is likewise reversed. The terms 'upper' and 'lower', when applied to limits of a definite integral, refer to their positions above and below the integral sign and have nothing to do with their values relative to one another.

Now look at Fig. 8.4(c) and (d), where the area lies in the negative range of y. If $b > a$, as in Fig. 8.4(c), the definite integral is negative, and if $b < a$, as in Fig. 8.4(d), the result is positive. These facts demonstrate further the difference between the purely mathematical concept of the definite integral, and its physical representation as the area under a curve. The latter must always be a positive quantity, whereas the former may be positive or negative. But an important practical consideration arises if the curve of $y = f(x)$ crosses the x-axis between the limits of integration. The total numerical area must then be evaluated as the sum of two separate areas — that above the x-axis and that below the x-axis — each area being assumed positive.

Area between two curves

Suppose it is required to find the shaded area in Fig. 8.5, i.e. the area between the curve of $y = f(x)$ and the curve of $y = \phi(x)$ between the limits of $x = a$ and $x = b$. Now the area under the curve of $y = f(x)$ is given by $\int_a^b f(x)dx$, and the area under the curve of $y = \phi(x)$ is given by $\int_a^b \phi(x)dx$. Clearly, the shaded area is equal to the area under the curve of $y = f(x)$

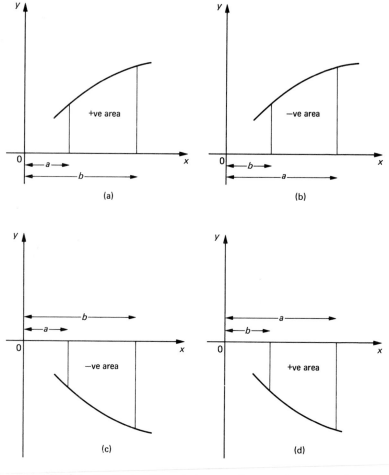

Fig. 8.4   'Positive' and 'negative' areas under curves (see page 145).

minus the area under the curve of $y = \phi(x)$, i.e.

$$S = \int_a^b f(x)\mathrm{d}x - \int_a^b \phi(x)\mathrm{d}x \qquad (8.12)$$

EXAMPLE 8.2
(a) Find the area enclosed by the curve $y = 2x - x^2$ and the straight line $y = 2$ between the limits of $x = 0$ and $x = 2$.
(b) Find the area enclosed by the curves $y = 4x - x^2$ and $y = x^2 - 2x + 2.5$

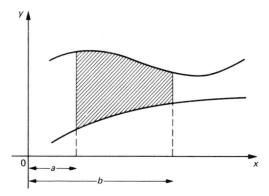

Fig. 8.5 The area between two curves, between the limits of $x = a$ and $x = b$. The upper curve is that of $y = f(x)$, and the lower curve depicts the function $y = \phi(x)$ (page 145).

(a) A diagram should always be drawn first, and the present situation is shown in Fig. 8.6(a), with the area to be evaluated shaded. Using 8.12, the area is given by

$$S = \int_0^2 2dx - \int_0^2 (2x - x^2)dx$$

$$= \left[2x - (x^2 - x^3/3) \right]_0^2$$

$$= [2(2) - \{(2)^2 - (2)^3/3\}] - [2(0) - \{(0)^2 - (0)^3/3\}]$$

$$= 4 - 4 + 8/3 = 2\tfrac{2}{3}$$

(b) The situation is shown in Fig. 8.6(b), and since no limits are specified, it is evident that the whole enclosed area between the points of intersection of the two curves is required, and so these must be found first. The points of intersection lie where both equations are satisfied together, and so are found by equating the right-hand sides of the equations of the two curves, giving

$$x^2 - 2x + 2.5 = 4x - x^2$$

i.e.

$$2x^2 - 6x + 2.5 = 0 \tag{8.13}$$

Solving 8.13 for x will yield the abscissae of the points of intersection, which is all we need to set limits on the integral. Using the formula for solving a quadratic equation (see equations 4.1 and 4.2), we have

$$x = \frac{6 \pm \sqrt{(36 - 20)}}{4} = \frac{6 \pm \sqrt{16}}{4} = \frac{6 \pm 4}{4}$$

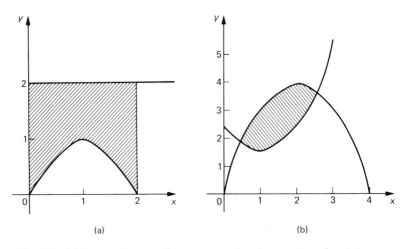

Fig. 8.6   (a) The area between the curve of the function $y = 2x - x^2$ and the straight line $y = 2$, between the limits of $x = 0$ and $x = 2$ (see Example 8.2(a)). (b) The area enclosed by the curves $y = 4x - x^2$ and $y = x^2 - 2x + 2.5$ (see Example 8.2(b)).

So either

$$x = \tfrac{1}{2} \quad \text{or} \quad x = 2\tfrac{1}{2}.$$

Hence the two curves intersect at the points whose x co-ordinates are $\tfrac{1}{2}$ and $2\tfrac{1}{2}$, and so these two values of x are the lower and upper limits of integration respectively. Therefore, the required area is given by

$$S = \int_{1/2}^{5/2} (4x - x^2)\mathrm{d}x - \int_{1/2}^{5/2} (x^2 - 2x + 2.5)\mathrm{d}x$$

$$= \left[2x^2 - x^3/3 - x^3/3 + x^2 - 2.5x \right]_{1/2}^{5/2}$$

$$= \left[3x^2 - 2x^3/3 - 5x/2 \right]_{1/2}^{5/2}$$

$$= \left\{ 3\left(\frac{5}{2}\right)^2 - 2\left(\frac{5}{2}\right)^3\Big/3 - 5\left(\frac{5}{2}\right)\Big/2 \right\} - \left\{ 3\left(\frac{1}{2}\right)^2 - 2\left(\frac{1}{2}\right)^3\Big/3 - 5\left(\frac{1}{2}\right)\Big/2 \right\}$$

$$= \frac{3(25)}{4} - \frac{2(125)}{3(8)} - \frac{25}{4} - \frac{3}{4} + \frac{2}{24} + \frac{5}{4} = \frac{482 - 418}{24} = \frac{64}{24} = \frac{8}{3} = 2\tfrac{2}{3}$$

Hence the area enclosed by the curves $y = 4x - x^2$ and $y = x^2 - 2x + 2.5$ is $2\frac{2}{3}$ square units.

At a first reading skip to page 152 for Numerical Integration.

INTEGRATION AS A SUMMATION

Consider the situation shown in Fig. 8.2, which illustrated the proof of the area under a curve. The proof involved dividing the area up into a large number of vertical strips and considering any one of these. A typical strip as shown in Fig. 8.2, PQNM, and its area is approximately $y \cdot \delta x$. Now the whole area ABCD is made up of the sum of the areas of all the vertical strips between $x = a$ and $x = b$. Thus

$$S \simeq \sum_{x=a}^{b} y \cdot \delta x = \sum_{x=a}^{b} f(x)\delta x$$

(see page 161 for details concerning the Σ-notation); or, since the area of each strip approaches $y \cdot \delta x$ more accurately as $\delta x \to 0$,

$$S = \lim_{\delta x \to 0} \left\{ \sum_{x=a}^{b} f(x)\delta x \right\}$$

But we have already shown that $S = \int_a^b f(x)\mathrm{d}x$, so

$$\int_a^b f(x)\mathrm{d}x = \lim_{\delta x \to 0} \left\{ \sum_{x=a}^{b} f(x)\mathrm{d}x \right\} \qquad (8.14)$$

Relationship 8.14 is very important, as it shows that if we have a situation wherein a very large number of very small elements need to be added together, the summation may be replaced by an integral.

EXAMPLE 8.3

Find the area of a circle of radius r.

Let the circle be dissected into a large number of very thin rings, each of width δx, as shown in Fig. 8.7. Consider a typical ring whose inner edge is distant x from the centre of the circle. The circumference of the inner edge of the ring is $2\pi x$. Since δx is small, the area of the ring is approximately $2\pi x \cdot \delta x$, as can be appreciated by imagining the ring to be cut out and straightened, when it would approximate to a rectangle of length $2\pi x$ and width δx. Hence the area of the circle, S, is equal to the sum of all

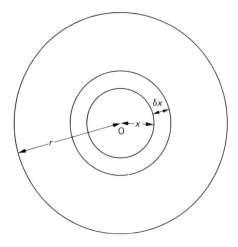

Fig. 8.7   A circle of radius r dissected into thin rings of width δx (see page 149).

such rings, i.e.

$$S = \lim_{\delta x \to 0}\left(\sum_{x=0}^{r} 2\pi x \cdot \delta x\right) = \int_0^r 2\pi x\,dx$$

$$= 2\pi \int_0^r x\,dx = 2\pi[x^2/2]_0^r$$

$$= \pi[x^2]_0^r = \pi r^2$$

The mean value of a function

The mean, or average, value of a set of n quantities: $y_1, y_2, y_3, \ldots, y_n$, is given by

$$\bar{y} = \frac{y_1 + y_2 + y_3 + \cdots + y_n}{n} = \frac{\Sigma y}{n}$$

Can we find the mean value of a function (of x) between two specified values of x? Look at Fig. 8.8 which shows a graph of the function of x, i.e. of $y = f(x)$, and suppose it is desired to find the mean value of $f(x)$ between the values of $x = a$ and $x = b$. Imagine the interval to be divided up into sub-intervals as shown, with each sub-interval being of width δx. The more sub-intervals there are, the smaller is δx, and as $n \to \infty$, so $\delta x \to 0$.

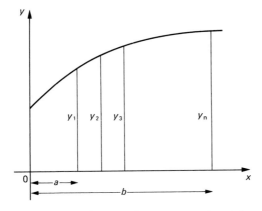

Fig. 8.8   Some ordinates between those at $x = a$ and $x = b$ (see page 150).

Now the mean value of the curve is given by

$$\bar{y} = \lim_{n \to \infty} \left(\frac{y_1 + y_2 + \cdots + y_n}{n} \right) = \lim_{n \to \infty} \left\{ \frac{(y_1 + y_2 + \cdots + y_n)\delta x}{n \cdot \delta x} \right\}$$

the last term above being obtained from the previous one by multiplying both numerator and denominator by δx. Hence

$$\bar{y} = \lim_{n \to \infty} \left(\frac{\sum\limits_{x=a}^{b} y \cdot \delta x}{n \cdot \delta x} \right)$$

Now since $y = f(x)$, $(n - 1)\delta x = b - a$, and $\delta x \to 0$ as $n \to \infty$

$$\bar{y} = \lim_{\delta x \to 0} \left\{ \frac{\sum\limits_{x=a}^{b} f(x)\delta x}{b - a + \delta x} \right\} = \frac{\lim\limits_{\delta x \to 0} \left\{ \sum\limits_{x=a}^{b} f(x)\delta x \right\}}{\lim\limits_{\delta x \to 0} \{b - a + \delta x\}}.$$

The last term follows because the limit of a quotient equals the quotient of two limits. Therefore

$$\bar{y}_i = \frac{\displaystyle\int_a^b f(x)\mathrm{d}x}{b - a} = \frac{\text{area under the curve}}{\text{interval of } x} \qquad \text{(page 149)}$$

which is usually written as

$$\bar{y} = \frac{1}{b-a} \int_a^b f(x)dx \qquad (8.15)$$

EXAMPLE 8.4
Find the mean value of the function x^2 between the limits of $x = 1$ and $x = 3$. We shall denote the mean value of a function $f(x)$ by $\overline{f(x)}$. Then

$$\overline{f(x)} = \frac{1}{3-1} \int_1^3 x^2 dx = \frac{1}{2}\left[\frac{x^3}{3}\right]_1^3 = 4\frac{1}{3}$$

Note that $f(1) = 1$ and $f(3) = 9$. but the mean value of $f(x)$ between limits $x = 1$ and $x = 3$ is $4\frac{1}{3}$. As with a mean value in statistics, the calculation has given due weight to the distribution of values of the function within the interval 1 to 3 (see Fig. 8.3).

NUMERICAL INTEGRATION

In the previous chapter it was stated that it was not possible to integrate certain functions. Now this does not mean to say that for such a function, $f(x)$, its integral, $\int f(x)dx$, does not exist. What it does mean is that for integration, which is defined as usual as

$$\int f(x)dx = F(x) + c$$

no known method will enable us to find $F(x)$.

However, if we put $y = f(x)$, then for most functions a graph could be drawn and an area under the curve of $y = f(x)$ between two limits $x = a$ and $x = b$ defined. Hence we know that the definite integral, $\int_a^b f(x)dx$, exists because the area just defined is the physical representation of the integral; but one is unable to evaluate the definite integral by direct analytical means.

As an example, consider the function 2^{x^2}; this cannot be integrated by any known method. However, the curve of $y = 2^{x^2}$ is shown in Fig. 8.9, and an area under the curve is defined between $x = 1$ and $x = 2$. This area is identified with the definite integral $\int_1^2 2^{x^2} dx$, but the latter cannot be used for evaluating the area. If, however, some other method could be used for at least *estimating* the shaded area, the result would also be an approximation of the value of the definite integral. Such procedures do exist, and are called methods of **numerical integration**: two will be described here.

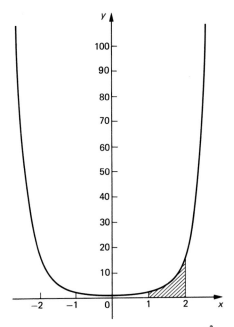

Fig. 8.9   Curve of the function $y = 2^{x^2}$.

Counting squares

This method is very elementary, and laborious. It consists of drawing a very accurate graph of the function over the range of x concerned, drawing the vertical lines at the limits, and then counting the squares of the graph paper within the required area. Assuming that the graph is drawn on metric paper, every 1 mm square would need to be counted. This would present no problems on the three straight sides of the area (apart from tedium!), but much subjectivity is required along the edge defined by the curve itself.

Simpson's Rule

This is the most well-known method of numerical integration; it is capable of giving good results, and is not too laborious to carry out. The rule will be stated below, but not proved.

Suppose we require to evaluate $\int_a^b f(x)dx$ by Simpson's Rule. First, consider a graph of $y = f(x)$ with the area marked out between the limits $x = a$

and $x = b$ (Fig. 8.10). Then divide the area into an *even* number of strips of width Δx (see page 78 for discussion of the notations Δx and δx). There is no question here of the strips being very narrow and very large in number; in Fig. 8.10 there are only 4 strips. Since a, b, and Δx are known, it follows that the x-values (abscissae) of the vertical lines defining the strips can be evaluated, and by substituting these abscissae into $f(x)$, the corresponding y-values (ordinates) can be calculated. Simpson's Rule then gives the area as

$$S \simeq \text{one-third the width of a strip} \begin{cases} \text{first ordinate + last ordinate +} \\ \text{twice the sum of the remaining odd} \\ \text{ordinates + four times the sum of} \\ \text{the even ordinates} \end{cases}$$

For the situation shown in Fig. 8.10, where there are four strips, the formula for Simpson's Rule is

$$S \simeq \frac{\Delta x}{3} \{y_1 + y_5 + 2y_3 + 4(y_2 + y_4)\} \tag{8.16}$$

EXAMPLE 8.5

Evaluate $\int_1^2 2^{x^2}$ by Simpson's Rule, using (a) 4 strips, (b) 10 strips.
(a) It is useful to calculate the ordinates by means of a table

x	x^2	$2^{x^2} = y$	$2y$ (where appropriate)	$4y$ (where appropriate)
1.00	1.000	2.000	—	—
1.25	1.563	2.954	—	11.816
1.50	2.250	4.757	9.514	—
1.75	3.063	8.354	—	32.416
2.00	4.000	16.000	—	—

Thus

$$S \simeq \frac{0.25}{3} \{(2.000 + 16.000) + 9.514) + (11.816 + 32.416)\}$$

$$\simeq \frac{0.25 \times 71.746}{3} = 5.97883$$

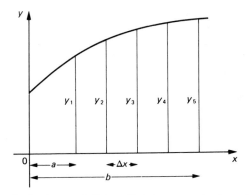

Fig. 8.10 Simpson's Rule to find the area under the curve.

(b)

x	x^2	$2^{x^2} = y$	$2y$ (where appropriate)	$4y$ (where appropriate)
1.0	1.00	2.000	—	—
1.1	1.21	2.313	—	9.252
1.2	1.44	2.713	5.426	—
1.3	1.69	3.227	—	12.908
1.4	1.96	3.890	7.780	—
1.5	2.25	4.757	—	19.028
1.6	2.56	5.895	11.790	—
1.7	2.89	7.143	—	29.652
1.8	3.24	9.445	18.890	—
1.9	3.61	12.190	—	48.760
2.0	4.00	16.000	—	—

Thus

$$S \simeq \frac{0.1}{3}\{2.000 + 16.000) + (5.426 + 7.780 + 11.790 + 18.890)$$
$$+ (9.252 + 12.908 + 19.028 + 29.652 + 48.760)\}$$
$$\simeq \frac{0.1 \times 181.486}{3} = 6.04953$$

Hence

$$\int_1^2 2^{x^2}\, dx \simeq 6.05$$

The more strips employed, the greater will be the accuracy of the result.

At a first reading, omit the remainder of this chapter.

INFINITE INTEGRALS

Sometimes, a definite integral may be evaluated even if one of the limits of the integrand is infinite. The usual definition, relationship 8.11, is modified slightly, and the modification differs according to which limit is infinite. Thus

$$\int_a^\infty f(x)dx = \lim_{x\to\infty} \{F(x) - F(a)\} \qquad (8.17)$$

$$\int_\infty^b f(x)dx = \lim_{x\to\infty} \{F(b) - F(x)\} \qquad (8.18)$$

where

$$d\{F(x)\}/dx = f(x)$$

It should be noted that if the limit does not exist, then the integral is meaningless and cannot be evaluated.

EXAMPLE 8.6

Evaluate, if possible,

(a) $\int_1^\infty \dfrac{dx}{x^2}$ (b) $\int_1^\infty \dfrac{dx}{x}$

(a) Using 8.17, we have

$$\int_1^\infty \frac{dx}{x^2} = \int_1^\infty x^{-2}dx = \left[-x^{-1}\right]_1^\infty = \left[-\frac{1}{x}\right]_1^\infty$$

$$= \lim_{x\to\infty} \left\{\left(-\frac{1}{x}\right) - (-1)\right\} = 1$$

(b) Again using 8.17, we have

$$\int_1^\infty \frac{dx}{x} = \left[\log_e x\right]_1^\infty \quad \text{(see page 00)}$$

$$= \lim_{x\to\infty} \{\log_e x - \log_e 1\}$$

Now as $x \to \infty$, $\log_e x \to \infty$. Hence the limit does not exist, and $\int_1^\infty dx/x$ is meaningless.

EXERCISES 8

1. Evaluate

(a) $\int_{1}^{3} (2x + x^2)dx$ (b) $\int_{4}^{9} \{(x + 1)/\sqrt{x}\}dx$

2. Evaluate $\int_{1}^{2} dx/x^2$ by the following methods.
(a) Draw an accurate graph of the functions $y = 1/x^2$ between $x = 1$ and $x = 2$, using as many intermediate x-values and as large a scale as practicable. Count the squares of the graph paper lying within the area enclosed by the curve, the x-axis, and the vertical lines $x = 1$ and $x = 2$. Express the results in square units.
(b) Use Simpson's Rule with 4 strips.
(c) Use Simpson's Rule with 10 strips.
(d) Integrate directly.
Compare the four results.

3*. Find the mean value of the following functions between the limits stated:
(a) $2x + x^2$ between $x = 1$ and 3,
(b) $(x + 1)/\sqrt{x}$ between $x = 4$ and 9.

4*. Evaluate, where possible:

(a) $\int_{1}^{\infty} dx/\sqrt{x}$ (b) $\int_{0}^{\infty} dx/(x + 2)^2$

9

Mathematical Series

A **series** is a succession of quantities which, after the first, are formed according to a common law. Each of the quantities is called a **term** of the series. For example:

3, 8, 13, 18, ... is a series, as each term is formed by adding 5 to the preceding term;

$a, a + d, a + 2d,$ ... is a series, since each term is formed from the previous term by adding d;

4, 12, 36, 108, ... is a series, since each term is formed by multiplying the preceding term by 3;

1, $2x$, $4x^2$, $8x^3$, ... is a series, each term formed by multiplying the previous term by $2x$.

Series are very important in mathematics. As biologists working through this book, our most immediate need of a series is to define e, the base of natural logarithms, which we shall do in the next chapter. But with biology coming to rely more on numerical methods than previously, the use of electronic computers is increasing in many fields of biology, and programming a computer often involves the use of series. You will already have met two kinds of series in your school-days, the arithmetic and the geometric; we shall revise the former briefly and develop some notation, and then go on to study the geometric series in greater depth.

THE ARITHMETIC SERIES

An **arithmetic series** is a series of terms that increase by a constant amount, which may be positive or negative. The following are examples of arithmetic series:

1, 2, 3, 4, . . .   common difference = +1;
$5/4$, $3/2$, $7/4$, 2, . . .   common difference = +¼;
6, 3, 0, −3, . . .   common difference = −3;
−2, −1½, −1, −½, . . . common difference = +½.

General notation

A general notation will now be developed for the arithmetic series, and a similar development may be made for any series. Consider an arithmetic series where the first term is a and whose common difference is d. Then

1st term $= a$
2nd term $= a + d$
3rd term $= (a + d) + d$ $= a + 2d$
4th term $= (a + 2d) + d = a + 3d$
$\vdots$ $\vdots$

ith term (general term) $= a + (i − 1)d$
$\vdots$ $\vdots$

nth term (last term) $= a + (n − 1)d$

The ith term is known as the **general term** because it indicates the general structure of any term in the series; any particular term can be evaluated from the given form of the ith term. For example, the 4th term of an arithmetic series whose first term, a, is 6, and whose common difference, d, is −3, is given by $6 + (4 − 1)(−3) = 6 + (3)(−3) = 6 − 9 = −3$. The nth term is by convention the **last term** in the series.

Sum of n terms

The most important feature of a series is the sum of a number of its terms, and it is very convenient to be able to find a formula for this sum. This is easy to do in the case of the arithmetic series.

Theorem 9.1. The sum, S_n, of n terms of an arithmetic series, whose first term is a and whose common difference is d, is given by

$$S_n = \frac{n}{2}\{2a + (n − 1)d\}$$

The last term of the series is $a + (n − 1)d$; hence

$$S_n = a + (a + d) + (a + 2d) + \cdots + a + (n − 1)d \qquad (9.1)$$

Reversing the series, we have

$$S_n = \{a + (n - 1)d\} + \{a + (n - 2)d\} + \cdots + a \tag{9.2}$$

Adding together 9.1 and 9.2 term by term, in order, gives

$$2S_n = [a + \{a + (n - 1)d\}] + [(a + d) + \{a + (n - 2)d\}]$$
$$+ [(a + 2d) + \{a + (n - 3)d\}] + \cdots + [\{a + (n - 1)d\}] + a \tag{9.3}$$

i.e.

$$2S_n = \{2a + (n - 1)d\} + \{2a + (n - 1)d\} + \cdots + \{2a + (n - 1)d\}$$

that is, there are n terms of the form $2a + (n - 1)d$. Thus

$$2S_n = n\{2a + (n - 1)d\}$$

or

$$S_n = \frac{n}{2}\{2a + (n - 1)d\} \tag{9.4}$$

EXAMPLE 9.1

In the series $5, 2, -1, \ldots$; what is

(a) the 15th term

(b) the sum of the first 15 terms, S_{15}?

(a) The first term, a, is 5; and the common difference, d, is -3. In the formula for the general term, $i = 15$. Hence the 15th term is $5 + (14)(-3) = -37$.

(b) Substituting in 9.4, with $n = 15$, we have

$$S_{15} = \frac{15}{2}\{(2)(5) + (14)(-3)\}$$

$$= \frac{(15)(-32)}{2} = -\frac{480}{2} = -240$$

EXAMPLE 9.2

Find the number of terms in the arithmetic series

(a) $8, 6, 4, \ldots$ and whose sum is 20;

(b) $5, 8, 11, \ldots$ and whose sum is 55.

(a) In this series, $a = 8$ and $d = -2$. From 9.4, we have

$$20 = \frac{n}{2}\{16 + (n - 1)(-2)\}$$

which reduces to the quadratic equation

$$n^2 - 9n + 20 = 0$$

i.e.

$$(n - 4)(n - 5) = 0$$

Therefore

$$n = 4 \quad \text{or} \quad n = 5.$$

It can be seen that either answer is correct since the first five terms are 8, 6, 4, 2, 0; and so either the first four or five terms sum to 20.

(b) Here, $a = 5$ and $d = 3$. From 9.4, we have

$$55 = \frac{n}{2}\{10 + (n - 1)(3)\}$$

which reduces to

$$3n^2 + 7n - 110 = 0$$

So

$$n = \frac{-7 \pm \sqrt{(49 + 1320)}}{6} \quad \frac{-7 \pm \sqrt{1369}}{6}$$

$$= \frac{-7 \pm 37}{6} = \frac{30}{6} \quad \text{or} \quad -\frac{44}{6}$$

Now n, the number of terms in a series, must be a positive integer, and so the answer of $-44/6$ is inadmissible. Hence, the number of terms in the series is 5.

Σ- AND Π-NOTATIONS

Σ-notation

The sum of the first 4 terms of an arithmetic series whose first term is a, and whose common difference is d, is given by

$$S_4 = a + (a + d) + (a + 2d) + (a + 3d) \tag{9.5}$$

The individual terms on the right-hand side take the form of the general term $a + (i - 1)d$ with i changing from 1 to 4. This applies even to the first two terms, which could be represented as $a + 0d$ and $a + 1d$. The right-hand side of 9.5 can be written much more neatly, using the expression for the general term of the series, as

$$S_4 = \sum_{i=1}^{4} \{a + (i-1)d\} \tag{9.6}$$

The sign Σ means 'the sum of', and it implies the sum of terms of the type shown to the right of the sign, in the braces { } in this case. The actual terms to be summed are created by substituting integer values of i between the two limits shown below and above the Σ sign. If this is done in the expression 9.6, we obtain the summation on the right-hand side of 9.5. Compare the Σ-notation with that of the definite integral, and read pages 149 to 150 of Chapter 8.

EXAMPLE 9.3
   (a) Express the following in Σ-notation:

$$x_1 + x_2 + x_3 + \cdots + x_n$$

   (b) Expand $\sum_{i=3}^{7} i x^{(i-1)}$

(a) This is simply $\sum_{i=1}^{n} x_i$

(b) $\sum_{i=3}^{7} i x^{(i-1)} = 3x^{(3-1)} + 4x^{(4-1)} + 5x^{(5-1)} + 6x^{(6-1)} + 7x^{(7-1)}$

$$= 3x^2 + 4x^3 + 5x^4 + 6x^5 + 7x^6$$

Π-notation

Just as a summation can be concisely represented by means of a Σ-sign, so also can a product be similarly represented. Consider 4!; it is

$$4! = 1 \times 2 \times 3 \times 4$$

which can be written as

$$4! = \prod_{i=1}^{4} i$$

where the symbol Π means 'product of' the terms of type shown immediately to the right of the sign. The product notation is, however, very much less common than that for summation.

THE GEOMETRIC SERIES

A **geometric series** is a series of terms that increases by a constant ratio, which may be positive or negative. The following are examples of geometric series:

2, 4, 8, 16, . . .   common ratio   = +2

½, ¼, ⅛, ¹/₁₆, . . .   common ratio   = +½

−⅓, 1, −3, 9, . . .   common ratio   = −3

−2, ½, −⅛, ¹/₃₂, . . . common ratio   = −¼

In a geometric series whose first term is a, and whose common ratio is r, the 2nd term is ar, the 3rd term is $(ar)r = ar^2$, etc.; hence the ith term is $ar^{(i-1)}$, and the last term is $ar^{(n-1)}$.

Sum of n terms

Consider a geometric series whose first term is a, and whose common ratio is r. In this series

$$S_n = a + ar + ar^2 + \cdots + ar^{(n-2)} + ar^{(n-1)} \tag{9.7}$$

Multiply both sides by r:

$$rS_n = ar + ar^2 + \cdots + ar^{(n-2)} + ar^{(n-1)} + ar^n \tag{9.8}$$

9.7 minus 9.8 gives

$$S_n - rS_n = a - ar^n$$

i.e.

$$S_n(1-r) = a(1-r^n)$$

Hence

$$S_n = \frac{a(1-r^n)}{1-r} \tag{9.9}$$

Equally well, 9.7 could have been subtracted from 9.8 above, then the final result would have been

$$S_n = \frac{a(r^n-1)}{r-1} \tag{9.10}$$

Expression 9.9 is used if the modulus of r (that is, the value of r regardless of sign) is less than 1, and 9.10 is used if the modulus of r is greater than 1. These results can be expressed concisely in the form of a theorem, as was done for the sum of an arithmetic series.

Theorem 9.2. The sum, S_n, of n terms of a geometric series, whose first term is a and whose common ratio is r, is given by

$$S_n = \frac{a(r^n - 1)}{r - 1} \quad |r| > 1 \qquad \textit{Divergent}$$

or by

$$S_n = \frac{a(1 - r^n)}{1 - r} \quad |r| < 1 \qquad \textit{Convergent}$$

EXAMPLE 9.4

In the series $3, -{}^3\!/_2, {}^3\!/_4, \ldots$; what is

(a) the 13th term
(b) the sum of the first 13 terms, S_{13}?

(a) The first term, a, is 3; and the common ratio, r, is $-\frac{1}{2}$. In the formula for the general term, $i = 13$. Hence the 13th term is

$$(3)(-\tfrac{1}{2})^{12} = \frac{3}{4096}$$

(b) The modulus of r, $|r|$, is less than 1, so to sum the first 13 terms of the series, we employ equation 9.9:

$$S_{13} = \frac{3\{1 - (-\tfrac{1}{2})^{13}\}}{1 - (-\tfrac{1}{2})} = \frac{3\{1 + 1/8192\}}{1 + \tfrac{1}{2}}$$

$$= \frac{(2)(8193)}{8192} = 2.0002 \text{ (correct to 4 decimal places)}$$

Sum to infinity: divergence and convergence

The sum to infinity of a geometric series means the sum of an infinite number of terms of the series. Now at first sight this may seem a strange, even impossible, thing to find; and even if it were possible, the sum would apparently be infinitely large, and so useless for any practical purpose. The last two statements of the previous sentence are true enough for the arithmetic series, but they are not necessarily true for the geometric series. In fact, the concept of the sum to infinity of a series is very important in mathematics, and we shall use the geometric series as an illustration.

First, consider the geometric series: $2, 4, 8, 16, \ldots$, in which the common ratio is 2. The sum of the first two terms is 6, of the first three 14, of the first four 30, and of the first five terms 62. Clearly, not only is S_n increasing as n increases, but the increase of S_n with each successive increment in n becomes steadily greater. Evidently, if an infinite number of terms were added together the sum would become infinitely large. Thus for the geometric series 2, 4, 8, 16, . . . we may write

$$S_n \to \infty, \quad \text{as} \quad n \to \infty$$

Such a series is said to be **divergent**.

Now consider the series 1, $\frac{1}{2}$, $\frac{1}{4}$, $\frac{1}{8}$, ... The sum of the first two terms is $1\frac{1}{2}$, of the first three $1\frac{3}{4}$, of the first four $1\frac{7}{8}$, and of the first five terms $1\frac{15}{16}$. In this case, although S_n is increasing with each term added, it seems to be doing so more and more slowly. We are thus led to ask for this series, 'is there a limit above which S_n cannot rise, no matter how large n is?' This question may be answered by the use of the formula for the sum of a geometric series; expression 9.9, since $r = \frac{1}{2}$ for this series, and $a = 1$. Substituting in 9.9, we have for the sum of n terms

$$S_n = \frac{1\{1 - (\frac{1}{2})^n\}}{1 - \frac{1}{2}} = 2\{1 - (\frac{1}{2})^n\} \tag{9.11}$$

Consider now what happens when n gets very large, in other words $\frac{1}{2}$ is raised to a large power. Recall that $(\frac{1}{2})^2 = \frac{1}{4}$, $(\frac{1}{2})^3 = \frac{1}{8}$, $(\frac{1}{2})^4 = \frac{1}{16}$, and so on. The larger n becomes, the smaller is $(\frac{1}{2})^n$; so that as

$$n \to \infty, \quad (\frac{1}{2})^n \to 0$$

Hence, from 9.11, we can write that as

$$n \to \infty, \quad S_n \to 2.$$

Thus the sum to infinity of the series 1, $\frac{1}{2}$, $\frac{1}{4}$, $\frac{1}{8}$, . . . is 2, a perfectly ordinary and usable number. The series is said to be **convergent** because, as more an more terms are added in, the value of the sum of the terms converges to a definite limit.

For a geometric series, the criterion for convergence or divergence is the value of the common ratio, r. If the modulus of the common ratio, $|r|$, is greater than unity, then the series is divergent, and if $|r|$ is less than unity, the series is convergent.

Theorem 9.3. The sum to infinity of a convergent geometric series is given by

$$S_\infty = \frac{a}{1 - r} \tag{9.12}$$

Since this series is stated to be convergent, which implies $|r| < 1$, S_∞ can be obtained from formula 9.9. It is immediately apparent that with $|r| < 1$ as $n \to \infty$, $r^n \to 0$. Hence, the result 9.12 is proved.

EXAMPLE 9.5

Sum to infinity the series 3, $-^3/_4$, $^3/_{16}$, ... In this series, $a = 3$ and $r = -^1/_4$. So, substituting into formula 9.12

$$S_\infty = \frac{3}{1 - (-^1/_4)} = \frac{12}{5} = 2.4$$

EXAMPLE 9.6

Convert the recurring decimal 15.2$\overset{..}{6}\overset{.}{4}$ into a (non-decimal) fraction (vulgar fraction).

$$15.2\overset{..}{6}\overset{.}{4} = 15 + \frac{2}{10} + \frac{64}{10^3} + \frac{64}{10^5} + \frac{64}{10^7} + \cdots$$

$$= 15 + \frac{2}{10} + \frac{64}{10^3}\left\{1 + \frac{1}{10^2} + \frac{1}{10^4} + \cdots\right\}$$

$$= 15 + \frac{2}{10} + \frac{64}{10^3} \cdot \frac{1}{1 - (^1/_{10})^2}$$

(summing the series in the braces, above, to infinity)

$$= 15 + \frac{2}{10} + \frac{64}{990} = 15\frac{262}{990}$$

At first sight, series may appear to be of intrinsic mathematical interest only, but Example 9.6 demonstrates a very practical application of the geometric series. Most series have practical applications in one way or another, which make them important to the user of mathematical methods.

At a first reading you may now proceed direct to Chapter 10.

THE BINOMIAL SERIES

By direct multiplication, we can show that

$$(a + x)^2 = a^2 + 2ax + x^2$$
$$(a + x)^3 = a^3 + 3a^2x + 3ax^2 + x^3$$
$$(a + x)^4 = a^4 + 4a^3x + 6a^2x^2 + 4ax^3 + x^4$$

and so on. These results can be generalized, and also extended to cover situations where the exponent, n, in the expression $(a + x)^n$ is any real number, by a result known as the **binomial theorem**. This theorem gives the following result:

$$(a + x)^n = a^n + na^{(n-1)}x + \frac{n(n-1)}{2!}a^{(n-2)}x^2 + \frac{n(n-1)(n-2)}{3!}a^{(n-3)}x^3$$

$$+ \cdots + \frac{n(n-1)(n-2)\ldots(n-i+2)}{(i-1)!}a^{(n-i+1)}x^{(i-1)} + \cdots$$

$$(9.13)$$

This relationship can be written more succinctly by a fuller use of the factorial notation:

$$(a + x)^n = a^n + na^{(n-1)}x + \frac{n!}{2!(n-2)!}a^{(n-2)}x^2 + \frac{n!}{3!(n-3)!}a^{(n-3)}x^3$$

$$+ \cdots + \frac{n!}{(i-1)!(n-i+1)!}a^{(n-i+1)}x^{(i-1)} + \cdots \qquad (9.14)$$

The last term written out in 9.13 and 9.14 is the ith term of the series, and the sum of the series on the right-hand side of these two expressions is, of course, the left-hand side, $(a + x)^n$. This is the reverse situation to that considered hitherto. When discussing the arithmetic and geometric series, we already had the series of terms and were interested in finding a concise formula for the sum of the terms. Now the question is posed in reverse: can we express the concise expression $(a + x)^n$ as a series of terms? Relationships 9.13 and 9.14 answer this question in the affirmative, but the general proof of the binomial theorem is outside the scope of this book.

EXAMPLE 9.7
  Expand

(a) $(a + x)^5$
(b) $(a - x)^4$
(c) $(2 - 3y)^4$

(a) This involves the direct use of 9.13:

$$(a + x)^5 = a^5 + 5a^4x + \frac{(5)(4)}{(1)(2)}a^3x^2 + \frac{(5)(4)(3)}{(1)(2)(3)}a^2x^3$$

$$+ \frac{(5)(4)(3)(2)}{(1)(2)(3)(4)}ax^4 + \frac{(5)(4)(3)(2)(1)}{(1)(2)(3)(4)(5)}x^5$$

$$(a + x)^5 = a^5 + 5a^4x + 10a^3x^2 + 10a^2x^3 + 5ax^4 + x^5$$

(b) Put x in 9.13 equal to $-x$. Then 9.13 becomes, with $n = 4$,

$$\{a + (-x)\}^4 = a^4 + 4a^3(-x) + \frac{(4)(3)}{(1)(2)}a^2(-x)^2 + \frac{(4)(3)(2)}{(1)(2)(3)}a(-x)^3$$
$$+ \frac{(4)(3)(2)(1)}{(1)(2)(3)(4)}(-x)^4$$

i.e.

$$(a - x)^4 = a^4 - 4a^3x + 6a^2x^2 - 4ax^3 + x^4$$

(c) In 9.13, put x equal to $-3y$, $a = 2$, and $n = 4$; then

$$\{2 + (-3y)\}^4 = (2)^4 + (4)(2)^3(-3y) + \frac{(4)(3)}{(1)(2)}(2)^2(-3y)^2$$
$$+ \frac{(4)(3)(2)}{(1)(2)(3)}(2)(-3y)^3 + \frac{(4)(3)(2)(1)}{(1)(2)(3)(4)}(-3y)^4$$
$$= 16 + 32(-3y) + 24(-3y)^2 + 8(-3y)^3 + (-3y)^4$$

i.e.

$$(2 - 3y)^4 = 16 - 96y + 216y^2 - 216y^3 + 81y^4$$

In the above examples where the power, n, is a positive integer, it should be noted that there are $n + 1$ terms in the expansion. In the expansion of $(a + x)^n$, the first term is a^n (i.e. $a^n x^0$), and the last term is x^n (i.e. $a^0 x^n$); also, in each term the sum of the powers of a and x is always equal to n.

Negative and fractional indices

In all cases where n is not a positive integer, the expansion of $(a + x)^n$ produces an infinite number of terms.

EXAMPLE 9.8
Find the first 4 terms in the expansions of

(a) $\dfrac{1}{(2 + x)^3}$ (b) $\sqrt{(4 + 5x)}$ (c) $\dfrac{1}{\sqrt[3]{(8 - 3x)}}$

(a) $\dfrac{1}{(2 + x)^3} = (2 + x)^{-3}.$

In 9.13, put $a = 2$ and $n = -3$, then

$$(2+x)^{-3} = (2)^{-3} + (-3)(2)^{-4}x + \frac{(-3)(-4)}{(1)(2)}(2)^{-5}x^2$$

$$+ \frac{(-3)(-4)(-5)}{(1)(2)(3)}(2)^{-6}x^3 + \cdots$$

$$= \frac{1}{8} + \frac{(-3)x}{16} + \frac{12x^2}{64} + \frac{(-60)x^3}{384} + \cdots$$

i.e.

$$(2+x)^{-3} = \frac{1}{8} - \frac{3}{16}x + \frac{3}{16}x^2 - \frac{5}{32}x^3 + \cdots$$

(b)
$$\sqrt{(4+5x)} = (4+5x)^{1/2}.$$

In 9.13, put $a = 4$, $x = 5x$, and $n = 1/2$. Then

$$(4+5x)^{1/2} = (4)^{1/2} + (1/2)(4)^{-1/2}(5x) + \frac{(1/2)(-1/2)}{(1)(2)}(4)^{-3/2}(5x)^2$$

$$+ \frac{(1/2)(-1/2)(-3/2)}{(1)(2)(3)}(4)^{-5/2}(5x)^3 + \cdots$$

$$= \sqrt{4} + \frac{5}{2\sqrt{4}}x + \frac{(25)(-1/4)}{2\sqrt{(4)^3}}x^2 + \frac{(125)(3/8)}{6\sqrt{(4)^5}}x^3 + \cdots$$

$$= 2 + \frac{5}{2\sqrt{4}}x - \frac{25}{8\sqrt{64}}x^2 + \frac{375}{48\sqrt{1024}}x^3 + \cdots$$

Hence

$$\sqrt{(4+5x)} = 2 + \frac{5}{4}x - \frac{25}{64}x^2 + \frac{125}{512}x^3 + \cdots$$

(c)
$$\frac{1}{\sqrt[3]{(8-3x)}} = \frac{1}{(8-3x)^{1/3}} = (8-3x)^{-1/3}$$

In 9.13, put $a = 8$, $x = -3x$, and $n = -1/3$. Then

$$\{8 + (-3x)\}^{-1/3} = (8)^{-1/3} + (-1/3)(8)^{-4/3}(-3x)$$

$$+ \frac{(-1/3)(-4/3)}{(1)(2)}(8)^{-7/3}(-3x)^2$$

$$+ \frac{(-1/3)(-4/3)(-7/3)}{(1)(2)(3)}(8)^{-10/3}(-3x)^3 + \cdots$$

$$= (8)^{-1/3} + \frac{(-1/3)(-3x)}{(8)^{4/3}} + \frac{(-1/3)(-4/3)(-3x)^2}{(1)(2)(8)^{7/3}}$$

$$+ \frac{(-1/3)(-4/3)(-7/3)(-3x)^3}{(1)(2)(3)(8)^{10/3}} + \cdots$$

$$= \frac{1}{(8)^{1/3}} + \frac{1}{(8)^{4/3}}x + \frac{4}{(2)(8)^{7/3}}x^2 + \frac{28}{(6)(8)^{10/3}}x^3 + \cdots$$

$$= \frac{1}{\sqrt[3]{8}} + \frac{1}{\sqrt[3]{(8)^4}}x + \frac{2}{\sqrt[3]{(8)^7}}x^2 + \frac{14}{3\sqrt[3]{(8)^{10}}}x^3 + \cdots$$

$$= \frac{1}{2} + \frac{1}{16}x + \frac{2}{128}x^2 + \frac{14}{(3)(1024)}x^3 + \cdots$$

Thus

$$\frac{1}{\sqrt[3]{(8 - 3x)}} = \frac{1}{2} + \frac{1}{16}x + \frac{1}{64}x^2 + \frac{7}{1536}x^3 + \cdots$$

Numerical uses of the binomial series

These are normally restricted to cases where x is small compared with a in $(a + x)^n$.

EXAMPLE 9.9
Evaluate

(a) $(1.98)^5$ (b) $\sqrt[3]{7.76}$

correct to 4 decimal places.

(a) $(1.98)^5 = (2 - 0.02)^5$.

In 9.13, put $a = 2$, $x = -0.02$, and $n = 5$. Then

$$(1.98)^5 = (2)^5 + (5)(2)^4(-0.02) + \frac{(5)(4)}{(1)(2)}(2)^3(-0.02)^2$$

$$+ \frac{(5)(4)(3)}{(1)(2)(3)}(2)^2(-0.02)^3 + \frac{(5)(4)(3)(2)}{(1)(2)(3)(4)}(2)(-0.02)^4 + (-0.02)^5$$

$$= 32 + (80)(-0.02) + (80)(0.0004) + (40)(-0.000008)$$

$$+ (10)(0.00000016) + (-0.0000000032)$$

$$= 32 - 1.6 + 0.032 - 0.00032 + 0.0000016 - 0.0000000032$$

Only the first 4 terms need be considered to obtain the answer correct to 4 decimal places, since terms 5 and 6 contain only zeros in the first 5 decimal places. Since, however, the power to which 1.98 is raised is a positive integer, the binomial series terminates and so an exact result may be obtained by using all the terms. Therefore, we have

$$(1.98)^5 = 30.4316815968 \quad \text{exactly,}$$

or

$$(1.98)^5 = 30.4317 \quad \text{correct to 4 decimal places.}$$

(b) $\sqrt[3]{7.76} = (8 - 0.24)^{1/3}$. In (9.13), put $a = 8$, $x = -0.24$, and $n = \frac{1}{3}$. Then

$$\sqrt[3]{7.76} = (8)^{1/3} + (1/3)(8)^{-2/3}(-0.24) + \frac{(1/3)(-2/3)}{(1)(2)}(8)^{-5/3}(-0.24)^2$$

$$+ \frac{(1/3)(-2/3)(-5/3)}{(1)(2)(3)}(8)^{-8/3}(-0.24)^3 + \ldots$$

$$= \sqrt[3]{8} + \frac{(-0.24)}{3\sqrt[3]{(8)^2}} + \frac{(-2)(-0.24)^2}{18\sqrt[3]{(8)^5}} + \frac{(10)(-0.24)^3}{162\sqrt[3]{(8)^8}} + \ldots$$

$$= 2 - \frac{0.24}{(3)(4)} - \frac{0.1152}{(18)(32)} - \frac{0.13824}{(162)(256)} + \ldots$$

$$= 2 - \frac{0.24}{12} - \frac{0.1152}{576} - \frac{0.13824}{41472} + \ldots$$

$$= 2 - 0.02 - 0.0002 - 0.000003 = 1.979796$$

Hence

$$\sqrt[3]{7.76} = 1.9798 \quad \text{correct to 4 decimal places.}$$

In general, an exact result is not to be expected for a root of any kind. However the decimal approximation may be obtained to any desired accuracy using as many terms in the series as necessary.

Example 9.9 demonstrates a practical use of the binomial series. In each part, the number to be raised to a power was split into two parts, the first being a whole number and the second being small compared with the first. The smaller the second term is compared with the first, the faster does the series converge; in other words, only a few terms will need to be utilized for a good approximation. In any case, the smaller term *must* be less than 1, otherwise there will be no convergence.

SOME OTHER IMPORTANT MATHEMATICAL SERIES

In the expansion of $(a + x)^n$, whatever the value of n, we found a series of terms which can be regarded as a polynomial of x, since successive terms contain successive powers of x. The binomial series polynomial has a finite number of terms only if n is a positive integer; otherwise there are an infinite number of terms. Many other important mathematical functions can be defined in terms of what might be called a polynomial of infinite degree. For instance, in Chapter 10, e^x is defined as

$$e^x = 1 + x + \frac{x^2}{2!} + \frac{x^3}{3!} + \dots$$

thus enabling e^x to be calculated for any value of x. Below are given series which enable the natural logarithm of a number to be computed, and also the trigonometrical functions of an angle when it is expressed in radians.

The logarithmic series

It is not possible to expand $\log_e x$ as an infinite series directly, but $\log_e(1 + x)$ can be so expressed when $|x| < 1$. The series is easily found by making use of the result

$$\log_e(1 + x) = \int (1 + x)^{-1} dx$$

where the constant of integration is set to zero (see page 192). Now, by the binomial theorem,

$$(1 + x)^{-1} = 1 - x + x^2 - x^3 + x^4 - x^5 + \dots$$

so

$$\log_e(1 + x) = \int (1 - x + x^2 - x^3 + \dots) dx$$

i.e.

$$\log_e(1 + x) = x - \frac{x^2}{2} + \frac{x^3}{3} - \frac{x^4}{4} + \dots \qquad (9.15)$$

Put $x = -x$ in 9.15, then

$$\log_e\{1 + (-x)\} = (-x) - \frac{(-x)^2}{2} + \frac{(-x)^3}{3} - \frac{(-x)^4}{4} + \dots$$

i.e.

$$\log_e(1 - x) = -x - \frac{x^2}{2} - \frac{x^3}{3} - \frac{x^4}{4} - \dots \qquad (9.16)$$

The relationships 9.15 and 9.16 are valid only when $|x| < 1$ (i.e. $-1 < x < 1$), because it is only when this condition holds that the right-hand sides converge. The nearer x is to zero, the faster the convergence. Hence the utility of these series in evaluating natural logarithms is very limited, to values near 1 in fact. Can we obtain from 9.15 and 9.16 a series which will enable the natural logarithm of *any* number to be evaluated? Put

$$y = \frac{1 + x}{1 - x} \qquad (9.17)$$

then by cross-multiplication and re-arrangement, we obtain

$$x = \frac{y - 1}{y + 1} \qquad (9.18)$$

From 9.18 it can be seen that whatever the value of y may be, x will always lie in the range $-1 < x < 1$. If, then, we can find a series for $\log_e y$ derived from 9.15 and 9.16, it will be valid for any value of y. Now from 9.17,

$$\log_e y = \log_e \left(\frac{1 + x}{1 - x} \right) = \log_e(1 + x) - \log_e(1 - x)$$

$$= \left(x - \frac{x^2}{2} + \frac{x^3}{3} - \frac{x^4}{4} + \dots \right) - \left(-x - \frac{x^2}{2} - \frac{x^3}{3} - \frac{x^4}{4} - \dots \right)$$

$$= x - \frac{x^2}{2} + \frac{x^3}{3} - \frac{x^4}{4} + \dots + x + \frac{x^2}{2} + \frac{x^3}{3} + \frac{x^4}{4} + \dots$$

$$= 2x + 2\frac{x^3}{3} + 2\frac{x^5}{5} + \dots \text{ (since all terms in even powers of } x \text{ cancel out)}$$

Hence

$$\log_e y = 2\left(x + \frac{x^3}{3} + \frac{x^5}{5} + \ldots\right)$$

But from 9.18, $x = (y - 1)/(y + 1)$ so

$$\log_e y = 2\left\{\frac{y - 1}{y + 1} + \frac{1}{3}\left(\frac{y - 1}{y + 1}\right)^3 + \frac{1}{5}\left(\frac{y - 1}{y + 1}\right)^5 + \ldots\right\} \quad (9.19)$$

EXAMPLE 9.10

Evaluate $\log_e 3$ correct to 4 decimal places. Put $y = 3$ in 9.19; then

$$\log_e 3 = 2\left\{\frac{3 - 1}{3 + 1} + \frac{1}{3}\left(\frac{3 - 1}{3 + 1}\right)^3 + \frac{1}{5}\left(\frac{3 - 1}{3 + 1}\right)^5 + \frac{1}{7}\left(\frac{3 - 1}{3 + 1}\right)^7 + \ldots\right\}$$

$$= 2\left\{\frac{2}{4} + \frac{1}{3}\left(\frac{2}{4}\right)^3 + \frac{1}{5}\left(\frac{2}{4}\right)^5 + \frac{1}{7}\left(\frac{2}{4}\right)^7 + \frac{1}{9}\left(\frac{2}{4}\right)^9 + \frac{1}{11}\left(\frac{2}{4}\right)^{11} + \ldots\right\}$$

$$= 2\left\{\frac{1}{2} + \left(\frac{1}{3}\right)\left(\frac{1}{8}\right) + \left(\frac{1}{5}\right)\left(\frac{1}{32}\right) + \left(\frac{1}{7}\right)\left(\frac{1}{128}\right) + \left(\frac{1}{9}\right)\left(\frac{1}{512}\right) + \left(\frac{1}{11}\right)\left(\frac{1}{2048}\right) + \ldots\right\}$$

$$= 2\left(\frac{1}{2} + \frac{1}{24} + \frac{1}{160} + \frac{1}{896} + \frac{1}{4608} + \frac{1}{22528} + \ldots\right)$$

$$= 2(0.5 + 0.04167 + 0.006250 + 0.001116 + 0.0002170$$

$$+ 0.00004439 + \ldots)$$

$$= 2 \times 0.5492973 = 1.09859478$$

It must not be thought that this gives the answer correct to 8 decimal places, since we have only used the first 6 terms in the series, and some further terms would have non-zero values in the 5th to 8th decimal places. Since, however, the 6th term contains only zeros in the first 4 decimal places, we are able to state correctly that

$$\log_e 3 = 1.0986 \quad \text{correct to 4 decimal places}$$

Theorem 9.1 $$\lim_{x \to 0}\{\log_e(1 + x)\} = x \quad (9.20)$$

From relationship 9.15 we see that, if x is very small, all terms on the right-hand side after the first are negligible. Hence the result is proved.

Trigonometric series

If an angle, x, is measured in radians, it is possible to obtain the trigonometric functions $\sin x$ and $\cos x$ as the sums of infinite series. This enables the trigonometric functions of any angle to be obtained, without the use of tables or diagrams. As for the logarithmic series, we invoke the aid of the calculus to derive the series.

In line with the binomial and logarithmic series, we similarly require the series for $\sin x$ and $\cos x$ to be polynomial in form: so assume that

$$\sin x = a_0 + a_1 x + a_2 x^2 + a_3 x^3 + a_4 x^4 + a_5 x^5 + \ldots$$

$$\cos x = b_0 + b_1 x + b_2 x^2 + b_3 x^3 + b_4 x^4 + b_5 x^5 + \ldots$$

We further know that $\sin 0 = 0$ and $\cos 0 = 1$, and so, putting $x = 0$ into the above series, we have $a_0 = 0$ and $b_0 = 1$. So the two series may be written:

$$\left.\begin{array}{l} \sin x = a_1 x + a_2 x^2 + a_3 x^3 + a_4 x^4 + a_5 x^5 + \ldots \\ \cos x = 1 + b_1 x + b_2 x^2 + b_3 x^3 + b_4 x^4 + b_5 x^5 + \ldots \end{array}\right\} \quad (9.21)$$

The fundamental results of the differential calculus of trigonometric functions (Chapter 12, pages 237 and 238) show that

$$\frac{d(\sin x)}{dx} = \cos x \quad \text{and} \quad \frac{d(\cos x)}{dx} = -\sin x \quad (9.22)$$

Differentiation of the first equation of 9.21, and using the first result of 9.22 gives

$$a_1 + 2a_2 x + 3a_3 x^2 + 4a_4 x^3 + 5a_5 x^4 + \ldots = 1 + b_1 x + b_2 x^2 + b_3 x^3 + b_4 x^4 + \ldots$$

Equating coefficients, since 9.22 holds for all x-values, we have

$$a_1 = 1, 2a_2 = b_1, 3a_3 = b_2, 4a_4 = b_3, 5a_5 = b_4, \text{etc} \quad (9.23)$$

Similarly, differentiation of the second equation of 9.21, and using the second result of 9.22 gives

$$b_1 + 2b_2 x + 3b_3 x^2 + 4b_4 x^3 + 5b_5 x^4 + \cdots$$
$$= -(a_1 x + a_2 x^2 + a_3 x^3 + a_4 x^4 + \cdots)$$

Equating coefficients:

$$b_1 = 0, 2b_2 = -a_1, 3b_3 = -a_2, 4b_4 = -a_3, 5b_5 = -a_4, \text{etc.} \quad (9.24)$$

Hence, combining the results of 9.23 and 9.24, we have

$$a_1 = 1$$

$$2a_2 = 0 \qquad a_2 = 0$$

$$3a_3 = -\frac{a_1}{2} \qquad a_3 = -\frac{a_1}{2 \times 3} \qquad\qquad = \frac{1}{3!}$$

$$4a_4 = -\frac{a_2}{3} \qquad a_4 = 0$$

$$5a = -\frac{a_3}{4} \qquad a_5 = -\frac{a_3}{4 \times 5} = \frac{a_1}{2 \times 3 \times 4 \times 5} = \frac{1}{5!} \quad \text{etc.}$$

So

$$\sin x = x - \frac{x^3}{3!} + \frac{x^5}{5!} - \frac{x^7}{7!} + \cdots$$

again, combining the results of 9.23 and 9.24, we have

$$b_1 = 0$$

$$2b_2 = -1 \qquad b_2 = -\frac{1}{2} \qquad = -\frac{1}{2!}$$

$$3b_3 = -\frac{b_1}{2} \qquad b_3 = 0$$

$$4b_4 = -\frac{b_2}{3} \qquad b_4 = -\frac{b_2}{3 \times 4} = \frac{1}{4!}$$

$$5b_5 = -\frac{b_3}{4} \qquad b_5 = -\frac{b_3}{4 \times 5} = 0 \quad \text{etc.}$$

So

$$\cos x = 1 - \frac{x^2}{2!} + \frac{x^4}{4!} - \frac{x^6}{6!} + \cdots$$

To use either of these series, substitute for x the angle in radians whose trigonometric function is required.

EXERCISES 9

1. Find the arithmetic series whose third term is -11, and whose seventh term is 5. What is the twentieth term of the series, and what is the sum of the first twenty terms?

2. Expand

(a) $\sum\limits_{i=1}^{5} x_i^2$ (b) $a \sum\limits_{i=0}^{3} x^i$ (c) $\sum\limits_{i=1}^{4} \dfrac{\sqrt{x_i}}{a + x_i}$ (d) $\prod\limits_{i=1}^{4} x_i$

3. Express the following in Σ- or Π-notation, as appropriate:

(a) $(x_1 + x_2 + x_3)/3$ (b) $2x^5 + x^4 - x^2 - 2x$ (c) $(x_1 . x_2 . x_3)/3$

(d) $\sqrt[3]{(x_1 . x_2 . x_3)}$

4. The sum to infinity of a geometric series is 36, and the second term of the series is 8. What is the series?

5. Convert to vulgar fractions:

(a) $0.\dot{1}$ (b) $0.\dot{2}\dot{7}$ (c) $0.\dot{1}2\dot{3}$

6.* Expand

(a) $(x + 1)^5$ (b) $(x - 1)^5$ (c) $(3x - 2y)^4$ (d) $(x^2 + 3y^2)^4$

7.* Expand, up to the fourth term:

(a) $1/(1 - x)$ (b) $\sqrt{(4 + x^2)}$ (c) $\sqrt[3]{(8 - x^3)^2}$

8.* Evaluate, correct to 5 decimal places:

(a) $\sqrt{1.02}$ (b) $\sqrt{2}$

9.* Evaluate, correct to 4 decimal places:

(a) $\log_e 2$ (b) $\log_e 0.8$ (c) $\sin 50°$ (d) $\cos 20°$

10

The Exponential and Related Functions of Biological Importance

THE EXPONENTIAL FUNCTION

In the discussion of the growth of a microbial culture (Chapter 6), a primary concept was that of the rate of increase of the number of organisms in an unlimited environment. Since under these conditions it is considered that each cell, on average, divides at a constant rate, then the rate of increase of the entire culture at any instant of time, t, will be proportional to the actual number of cells, n, present at that instant; i.e.

$$\frac{dn}{dt} = kn \tag{10.1}$$

where k is the constant of proportionality. Because 10.1 contains the term dn/dt, the expression is a differential equation, and we have already remarked (page 109) that a differential equation of the form of 10.1 can be 'solved' to obtain a function of the form

$$n = f(t) \tag{10.2}$$

which gives the number of organisms present at time t.

If equations 10.1 and 10.2 are cast into more general $x-y$ terms, we have

$$\frac{dy}{dx} = ky \tag{10.3}$$

and

$$y = f(x) \qquad (10.4)$$

Now 10.4 is a generalized way of writing down functions, and includes those that we have already studied, such as polynomials; evidently the form of equation in 10.3, which we have hitherto only met in connexion with our culture growth problems, can give rise to, or can be obtained from, equation 10.4.

Now put $k = 1$ in 10.3; then we have

$$\frac{dy}{dx} = y \qquad (10.5)$$

In words, 'the rate of change of y with respect to x equals y'. If we substitute for y in 10.5 using 10.4, we have

$$\frac{dy}{dx} = f(x) \qquad (10.6)$$

So we are implying that there is a function of the form $y = f(x)$ whose differential coefficient is the same as the function itself, i.e. that $dy/dx = f(x)$. Put in another way, we are saying that for this particular function of x, the rate of change of $f(x)$ with respect to x is always equal to $f(x)$ for any specified value of x. Precisely what form does $f(x)$ take for it to have this property?

Consider the third degree polynomial function $y = x^3$, where $dy/dx = 3x^2$. Draw-up a small table of values of y and of dy/dx for a few values of x:

x	y	$\frac{dy}{dx}$
0	0	0
1	1	3
2	8	12
3	27	27
4	64	48

We see that for this function in the range of x selected, y and dy/dx are equal at two values of x, namely $x = 0$ and $x = 3$. Between these two values of x the gradient of the curve $y = x^3$ exceeds the value of y itself, i.e. $dy/dx > y$, and when $x > 3$ the gradient is less than y, i.e. $dy/dx < y$. It can be stated now that no polynomial function can fulfil our present requirements that $dy/dx = y$ *always*.

The number e and the exponential series

If attention is turned to that other function with which we are familiar, namely $y = ac^x$, then $dy/dx = ac^x \cdot \log_e c$ (page 84). Clearly, the differential coefficient is rather similar to the original, differing only by the inclusion of a constant multiplier, $\log_e c$. Now the logarithm of a number to its own base always equals 1, e.g. $\log_{10} 10 = 1$ (see Exercise 6(a) at the end of Chapter 2); so that if $c = e$, we have $y = ae^x$, and $dy/dx = ae^x \cdot \log_e e = ae^x$. So, what is the value of e?

As already remarked, the fundamental property of e is that if we consider the function $y = e^x$, then dy/dx also equals e^x. Dispensing with the variable y, we can write

$$\frac{d(e^x)}{dx} = e^x \tag{10.7}$$

Relationship 10.7 can be said to define e^x, and hence e by putting $x = 1$, but in itself does not give us the numerical value of e. What we now ask is, can we find some more elementary function of x which can be differentiated from first principles, and when so differentiated it remains unchanged? If we can find such a function of x, then it must equal e^x.

The exponential series

Consider the series of terms

$$1 + x + \frac{x^2}{2!} + \frac{x^3}{3!} + \frac{x^4}{4!} + \cdots + \frac{x^i}{i!} + \cdots \tag{10.8}$$

This series is convergent (page 165) because in successive terms $i!$ increases faster than x^i. If x is small convergence is rapid, but if x is large the series converges slowly; but the main point is that the further to the right one goes in 10.8, the smaller the terms become. Although the first two terms of the series appear to be different from the remainder, we can bring them into line by writing the second term, x, as $x^1/1!$ and the first term, 1, as $x^0/0!$ (remembering that any number raised to the power zero is 1, and that $0! = 1$; page 11). So the series may be written succinctly as

$$\sum_{i=0}^{\infty} \frac{x^i}{i!}$$

implying the sum to infinity of the series.

Now consider the differentiation of 10.8. Close inspection shows the function to be a polynomial (of infinite degree), so we can differentiate it term by term. Let us put the whole series equal to $f(x)$, as clearly 10.8 is a function of x, despite being written as a series. Writing out the first few terms, we have

$$f(x) = 1 + x + \frac{x^2}{2!} + \frac{x^3}{3!} + \frac{x^4}{4!} + \frac{x^5}{5!} + \cdots \qquad (10.9)$$

then

$$f'(x) = \quad 1 + \frac{2x}{2!} + \frac{3x^2}{3!} + \frac{4x^3}{4!} + \frac{5x^4}{5!} + \cdots$$

and on cancelling in the numerator and denominator of each term, we find

$$f'(x) = 1 + x + \frac{x^2}{2!} + \frac{x^3}{3!} + \frac{x^4}{4!} + \cdots \qquad (10.10)$$

You will notice that the series 10.9 and 10.10 are exactly the same. True, all the terms have been shifted one place to the right, but because the series is infinite, the final result is unchanged. Hence $f(x) = f'(x)$, and so the series must be equal to e^x. Therefore

$$e^x = 1 + x + \frac{x^2}{2!} + \frac{x^3}{3!} + \cdots = \sum_{i=0}^{\infty} \frac{x^i}{i!} \qquad (10.11)$$

and so both relationships 10.7 and 10.11 define e^x.

The value of e

Now it is quite simple to evaluate e. All we need to do is put $x = 1$ in 10.11:

$$e = 1 + 1 + \frac{1}{2!} + \frac{1}{3!} + \cdots = \sum_{i=1}^{\infty} \frac{1}{(i-1)!} \qquad (10.12)$$

Consider the first 10 terms, in the following table:

i	$i - 1$	$(i-1)!$	$1/(i-1)!$
1	0	1	1.000 000
2	1	1	1.000 000
3	2	2	0.500 000
4	3	6	0.166 667
5	4	24	0.041 667
6	5	120	0.008 333
7	6	720	0.001 389
8	7	5 040	0.000 189
9	8	40 320	0.000 025
10	9	362 880	0.000 003

$$\sum_{i=1}^{10} \frac{1}{(i-1)!} = 2.718\ 282$$

Hence, by considering the first 10 terms in the exponential series, we find that e = 2.71828, correct to 5 decimal places. If a more accurate evaluation of e were required, that is, correct to more decimal places, all that need be done is to add in more terms to the series. By now it should be apparent why e is irrational, since it is equal to the sum to infinity of a series of terms.

EXAMPLE 10.1

Evaluate

(a) $e^{0.6}$ (b) $e^{-0.6}$

correct to 3 decimal places.

(a) Put $x = 0.6$ in 10.15, then

$$e^{0.6} = 1 + 0.6 + \frac{(0.6)^2}{2!} + \frac{(0.6)^3}{3!} + \frac{(0.6)^4}{4!} + \frac{(0.6)^5}{5!} + \cdots$$

$$= 1 + 0.6 + \frac{0.36}{2} + \frac{0.216}{6} + \frac{0.1296}{24} + \frac{0.07776}{120} + \cdots$$

$$= 1 + 0.6 + 0.18 + 0.036 + 0.0054 + 0.000648 + \cdots$$

$$= 1.822048$$

Therefore, $e^{0.6} = 1.822$, correct to 3 decimal places.

(b) $e^{-0.6}$ can be obtained either through

$$\frac{1}{e^{0.6}} = \frac{1}{1.822} = 0.5488$$

$$= 0.549 \text{ correct to 3 decimal places.}$$

or by putting $x = -0.6$ in 10.15, then

$$e^{-0.6} = 1 + (-0.6) + \frac{(-0.6)^2}{2!} + \frac{(-0.6)^3}{3!} + \frac{(-0.6)^4}{4!} + \frac{(-0.6)^5}{5!} + \cdots$$

$$= 1 - 0.6 + \frac{0.36}{2} - \frac{0.216}{6} + \frac{0.1296}{24} - \frac{0.07776}{120} + \cdots$$

$$= 1 - 0.6 + 0.18 - 0.036 + 0.0054 - 0.000648 = 0.548752$$

Therefore, $e^{-0.6} = 0.549$, correct to 3 decimal places.

It is often necessary to obtain e^x for a given value of x. This should never be done by writing a numerical approximation for e and then raising it to the required power, e.g. $2.71828^{1.3}$. Either a similar calculation to those shown in Example 10.1 must be undertaken, which is laborious, or tables must be used. Sometimes, full tables of e^x are available in which, for a given x, e^x can be obtained directly. The table of e^x at the back of this book is in standard format, and is used in a similar way to any other mathematical table.

If a full table of e^x is not available (and such a table is not always found in books of tables), one may use the fact that e^x is the anti-natural logarithm, and so e^x can be evaluated by using a table of natural logarithms inversely, that is by finding x in the body of the table and reading the corresponding value of e^x in the table margins.

EXAMPLE 10.2
Find

(a) e^4 (b) $e^{0.07}$

using a table of natural logarithms. Check your results in the table of e^x.

(a) From the subsidiary table at the bottom of the page of the table of natural logarithms, we see that 4 lies between 2.3026 ($= \log_e 10$) and 4.6052 ($= \log_e 100$); so $10 < e^4 < 100$. Now $4 - 2.3026 = 1.6974$, so 1.6974 is sought in the body of the main table; the margins of the table corresponding to 1.6974 are 5.46. But since the result lies between 10 and 100,

$$e^4 \simeq 54.6$$

(b) In the body of the table, we find that 0.07 lies between 0.0677 ($= \log_e 1.07$) and 0.0770 ($= \log_e 1.08$). Now $0.07 - 0.0677 = 0.0023$, which means that the column containing 23 is required on the extreme right of the table. The nearest approach is 20 in the column headed 2; so

$$e^{0.07} \simeq 1.072$$

Regardless of the table used (e^x or natural logarithms), e^{-x} can always be obtained by taking the reciprocal of e^x. For example,

$$e^{-4} \simeq 1/54.6 \simeq 0.0183 \quad \text{and} \quad e^{-0.07} \simeq 1/1.072 \simeq 0.9329.$$

The exponential function

If

$$y = e^x \tag{10.13}$$

the curve of this function will have an intercept on the y-axis of 1.

More generally, if $y = ae^x$ the intercept is a. We also know that

$$\frac{dy}{dx} = ae^x$$

or   (10.14)

$$\frac{dy}{dx} = y$$

The second equation in 10.14 is the premise with which we started, namely, 'that the rate of change of y with respect to x is equal to y itself'. Now it would be very rare to find that the rate of change would actually *equal y*, but instead it would be proportional, i.e.

$$\frac{dy}{dx} = by$$

where b is the constant of proportionality. If this is the case the first equation of 10.14 becomes

$$y = ae^{bx}$$ (10.15)

and 10.15 is called the **exponential function**. Strictly, $y = e^x$ is *the* exponential function, but 10.15 is a more general, and useful form.

If we take natural logarithms of both sides of 10.15, we have

$$\log_e y = \log_e a + bx \cdot \log_e e$$

and since $\log_e e = 1$, the above becomes

$$\log_e y = \log_e a + bx$$ (10.16)

Equation 10.16 represents a linear relationship between $\log_e y$ and x with a gradient of b and an intercept of $\log_e a$. Graphs are shown in Fig. 10.1 of equation 10.15, (a) and (b), and of equation 10.16, (c) and (d). In (a) and (c), are shown the curve and line when b takes positive values, and in (b) and (d) those when b is negative.

It will be noticed that there is a strong similarity between the graphs in Fig. 10.1 and those in Fig. 4.5. Also, if one compares equation 10.15 with 4.8, and 10.16 with 4.9 it is evident that there is great similarity. In the case of the latter pair, we have

$$\log_e y = \log_e a + bx$$

and

$$\log_e y = \log_e a + (\log_e c)x$$

so evidently b of the exponential function equals $\log_e c$ of the (supposedly)

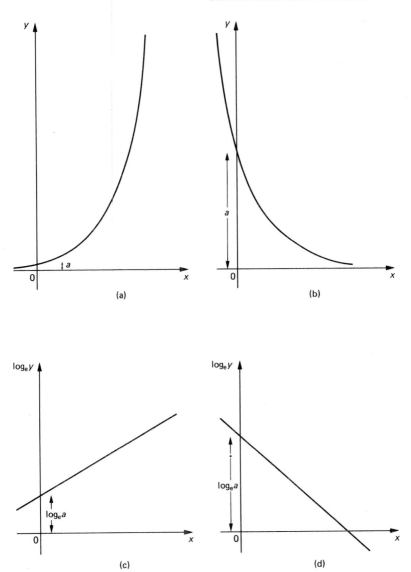

Fig. 10.1   The function $y = ae^{bx}$: (a) curve of the function when $b > 1$, (b) curve of the function when $b < 1$, (c) the straight line $\log_e y = \log_e a + bx$ corresponding to (a), (d) the straight line corresponding to (b).

more general function. Hence for any function of the form $y = ac^x$, there is a corresponding function, $y = ae^{bx}$, and the relationship between them is $b = \log_e c$. As the exponential function is much more generally useful, especially to the biologist, we shall now forget about functions of the form $y = ac^x$ altogether.

The exponential function is widely used in biology. With a positive value of b, the use of the function is confined to describing a particular form of growth (page 195), but the function in which b is negative has wider application. The decay of a radioactive element with time (of outstanding importance to the biologist interested in physiology and metabolism) proceeds according to the exponential function, and this function also describes the way that the intensity of a beam of light is attenuated as it passes through a liquid (Beer's Law). Empirically, the exponential function may also describe the attenuation of light through a foliage canopy and also the natural decline in numbers of individual plants in a cohort from seedling emergence to maturity. It is a matter of common observation that if a large number of seedlings of a particular species are found on a piece of bare ground, only a small proportion of these will reach maturity, and the decline in number of individuals with time may often be described by the exponential function. The example below illustrates this idea, as well as showing how the function is manipulated in calculations.

EXAMPLE 10.3

In a cohort of foxglove (*Digitalis purpurea*) plants, the number, y, surviving at time t (measured in months from the emergence of the seedlings) was found to conform to the equation

$$y = 100e^{-0.2310t}.$$

(a) What was the original number of seedlings in the cohort?
(b) What was the half-life of the cohort, i.e. the time when half the original number of individuals had died?
(c) The foxglove is a biennial, germinating in spring and flowering in the summer of the following year. Assuming that 15 months are required for vegetative growth, how many individuals of the cohort would be likely to survive to flower?
(a) Since time is measured from seedling emergence, we put $t = 0$; then, from the equation, $y = 100$. So the original number of seedlings was 100.
(b) To calculate the time at which half the number of seedlings have died, we substitute the number surviving into the equation for y and solve for t. In this case, half the original number surviving gives $y = 50$, and so

$$50 = 100e^{-0.2310t}$$

or

$$0.5 = e^{-0.2310t}$$

Taking the reciprocal of both sides gives

$$e^{0.2310t} = 2$$

Take natural logarithms of both sides

$$0.2310t = \log_e 2$$

Now

$$\log_e 2 = 0.6931,$$

and so

$$t = \frac{0.6931}{0.2310} \simeq 3$$

Therefore 50% of the plants die during the first three months from seedling emergence.

(c) As we are told that 15 months are required for vegetative growth, we merely substitute $t = 15$ into the equation and solve for y.

$$y = 100e^{-0.2310 \times 15}$$

i.e.

$$y = 100e^{-3.465}$$

or

$$y = 100/e^{3.465}$$

Now $e^{3.465} = 31.98$, and so

$$y \simeq 3.127$$

Hence, 3 plants are likely to survive to flower.

It should be noted that an exponential function, such as the one shown in the above example, would be fitted to data in its linear form, equation 10.16. In the foxglove cohort, observations on the number of individuals surviving would be made from time to time, and then a graph would be plotted of $\log_e$ (number of individuals) against time. The points should lie roughly on a straight line, and such a line may then be fitted in the way mentioned in Chapter 3 (page 51) and values of the constants $\log_e a$ (and hence a) and b obtained.

THE LOGARITHMIC FUNCTION

In the simplified form of the exponential function, where both the constants a and b are unity, we have $y = e^x$. Referring to equations 2.12 and 2.13, and the definition of a logarithm (page 14), we see that this implies

$$x = \log_e y \tag{10.17}$$

If we now reverse the roles of x and y, 10.17 becomes

$$y = \log_e x$$

A more generalized form of this is

$$y = \log_e (ax^b) \tag{10.18}$$

Expanding the right-hand side gives

$$y = \log_e a + \log_e x^b$$

or

$$y = \log_e a + b \cdot \log_e x \tag{10.19}$$

Relationship 10.19 shows y to be a linear function of $\log_e x$, the reverse situation to the exponential function in which $\log_e y$ is a linear function of x.

Graphs of the **logarithmic function** are shown in Fig. 10.2(a) and (b), illustrating the ordinary form (equation 10.18), and (c) and (d), illustrating the linear form (equation 10.19). In (a) and (c) of Fig. 10.2, b is positive, while in (b) and (d) this constant is negative.

The logarithmic function may describe empirically the relationship between the concentration, x, of a substance having a specific physiological action on an organism (e.g. a plant growth hormone) and a measurement, y, of the response of the organism (e.g. growth made) to the substance. If the logarithm of the concentration of the applied substance is plotted against the measurement of the response of the organism, and an approximately linear distribution of points is obtained, then the logarithmic function adequately describes the data, and the values of the constants a and b may be estimated. It is then possible to assess the concentration at which no response occurs, by observing the intercept of the curve *on the x-axis*, which is given by

$$x_0 = e^{-(\log_e a)/b}$$

(see Fig. 10.2 a).

In this example, the portion of the curve below the x-axis is biologically meaningless, since it would be most unlikely that very low concentrations of the physiologically active substance would induce a negative response, when at higher concentrations a positive response is obtained. Also, at very high concentrations, there would be a maximum response by the organism, and this would be approached asymptotically; but the logarithmic function has no upper asymptote at any value of y. For these reasons, the logarithmic function can only provide an empirical description *within the range of the data*, that is, only over the range of concentrations of the physiologically active substance included in the experiment.

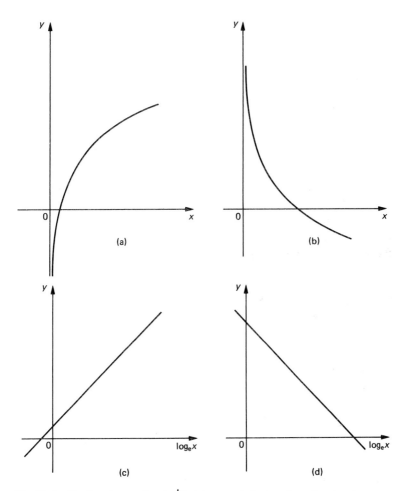

Fig. 10.2   The function $y = \log_e(ax^b)$: (a) curve of the function when $b > 1$, (b) curve of the function when $b < 1$, (c) the straight line $y = \log_e a + b.\log_e x$ corresponding to (a), (d) the straight line corresponding to (b).

THE CALCULUS OF EXPONENTIAL AND LOGARITHMIC FUNCTIONS

Basic results

Since $d(e^x)/dx = e^x$, then $\int e^x dx = e^x + c$. For the simple logarithmic function $y = \log_e x$, using the definition of a logarithm,

$$x = e^y \qquad (10.20)$$

Now

$$\frac{dx}{dy} = e^y$$

and so

$$\frac{dy}{dx} = \frac{1}{e^y}$$

But $e^y = x$, 10.20, so

$$\frac{dy}{dx} = \frac{1}{x}$$

Hence $d(\log_e x)/dx = 1/x$. This means that $\int dx/x = \log_e x + c$. These are very important results, and are summarized below:

$$\frac{d(e^x)}{dx} = e^x \qquad (10.21)$$

$$\frac{d(\log_e x)}{dx} = \frac{1}{x} \qquad (10.22)$$

$$\int e^x dx = e^x + c \qquad (10.23)$$

$$\int \frac{dx}{x} = \log_e x + c \qquad (10.24)$$

Equations 10.21 and 10.22 are additional basic rules for differentiation, while 10.23 and 10.24 are standard integrals.

It will be recalled that the standard integral first introduced, equation 7.6 on page 125, was not valid for $n = -1$, but was valid for any other value of n. The standard integral 10.24 covers the case where $n = -1$; i.e.

$$\int x^{-1} dx = \log_e x + c.$$

Some more complicated functions

Consider the exponential function, $y = e^{bx}$, i.e. where the coefficient a (equation 10.15) is 1. This can be differentiated by the function of a function rule. Put

$$u = bx \quad \text{then} \quad y = e^u \quad (\text{see } 10.20)$$

$$\frac{du}{dx} = b \quad \text{and} \quad \frac{dy}{du} = e^u$$

So

$$\frac{dy}{dx} = \frac{dy}{du} \cdot \frac{du}{dx} = be^u = be^{bx}$$

Hence

$$\frac{d(e^{bx})}{dx} = be^{bx} \tag{10.25}$$

From 10.25 we can say that $(\text{see } 10.24)$

$$\int be^{bx}dx = e^{bx} + c$$

i.e.

$$b \int e^{bx}dx = e^{bx} + c$$

Dividing both sides by b, we have

$$\int e^{bx}dx = \frac{1}{b} e^{bx} + k \qquad = \frac{c}{b} \tag{10.26}$$

where $k = c/b$. Thus expressions 10.25 and 10.26 give the differential coefficient and integral, respectively, of the function e^{bx}.

Now consider a different form of the logarithmic function,

$$y = \log_e(a + bx)$$

Again, this may be differentiated by the function of a function rule. Put

$$u = a + bx \quad \text{then} \quad y = \log_e u$$

$$\frac{du}{dx} = b \quad \text{and} \quad \frac{dy}{du} = \frac{1}{u} \quad \text{from } 10.22$$

So

$$\frac{dy}{dx} = \frac{dy}{du} \cdot \frac{du}{dx} = b\frac{1}{u} = \frac{b}{a + bx}$$

Hence

$$\frac{d\{\log_e(a + bx)\}}{dx} = \frac{b}{a + bx} \tag{10.27}$$

From 10.27 we can say that

$$\int \frac{b\,dx}{a + bx} = \log_e(a + bx) + c$$

i.e.

$$b \int \frac{dx}{a + bx} = \log_e(a + bx) + c$$

Dividing both sides by b, we have

$$\int \frac{dx}{a + bx} = \frac{1}{b} \cdot \log_e(a + bx) + k \tag{10.28}$$

where $k = c/b$. Thus expressions 10.27 and 10.28 give the differential coefficient and integral of $\log_e(a + bx)$ and $1/(a + bx)$, respectively.

EXAMPLE 10.4
　Find

(a) $d\{e^{(2+3x)}\}/dx$　　(b) $\int dx/(2 - 5x)$

(a) Use the function of a function rule.
Put

$$y = e^{(2+3x)} \quad \text{and} \quad u = 2 + 3x, \quad \text{then} \quad y = e^u$$

$$\frac{du}{dx} = 3 \quad \text{and} \quad \frac{dy}{du} = e^u$$

So

$$\frac{dy}{dx} = \frac{dy}{du} \cdot \frac{du}{dx} = 3e^u = 3e^{(2+3x)}$$

Hence

$$d\{e^{(2+3x)}\}/dx = 3e^{(2+3x)}$$

(b) This is a direct application of the standard integral, 10.28, where $a = 2$ and $b = -5$. Hence

$$\int \frac{dx}{2 - 5x} = -\frac{1}{5} \cdot \log_e(2 - 5x) + c$$

EXAMPLE 10.5*
   Find

(a) $\int(\log_e x)dx$ (b) $\int\{\log_e(a + bx)\}dx$

(a) Even this simple function is not straightforward to integrate; it has to be integrated by parts. It is an example, referred to at the end of Chapter 7 (page 137), in which the method of integration by parts is applied to a function, rather than to a product of two functions. In the formula for integration by parts 7.19, put $\psi(x) = \log_e x$ and $\phi(x) = 1$. Now 7.19 becomes

$$\int (\log_e x)dx = (\log_e x)\int dx - \int \left\{ \frac{d(\log_e x)}{dx} \int dx \right\} dx + c$$

$$(10.29)$$

Remembering that $\int dx = x$ and $d(\log_e x)/dx = 1/x$, 10.29 becomes

$$\int (\log_e x)dx = x \cdot \log_e x - \int (1/x)x dx + c$$
$$= x \cdot \log_e x - \int dx + c$$
$$= x \cdot \log_e x - x + c$$

Thus

$$\int(\log_e x)dx = x(\log_e x - 1) + c \qquad (10.30)$$

(b) In 7.19, put $\psi(x) = \log_e(a + bx)$ and $\phi(x) = 1$; then

$$\int\{\log_e(a + bx)\}dx = \log_e(a + bx)\int dx - \int \left[\frac{d\{\log_e(a + bx)\}}{dx} \int dx \right] dx + c$$

$$(10.31)$$

Now

$$\int dx = x \quad \text{and} \quad d\{\log_e(a + bx)\}/dx = b/(a + bx),$$

so 10.31 becomes

$$\int\{\log_e(a + bx)\}dx = x \cdot \log_e(a + bx) - \int \frac{bx\,dx}{a + bx} + c \qquad (10.32)$$

The remaining integral, on the right-hand side, will have to be found by the method of substitution. Put

$$u = a + bx,$$

then

$$x = (u - a)/b,$$

and

$$dx = du/b.$$

So

$$\int \frac{bx\,dx}{a + bx} = \int \frac{b}{u} \cdot \frac{(u - a)}{b} \cdot \frac{du}{b} = \int \frac{(u - a)du}{bu}$$

$$= \frac{1}{b} \int \left(1 - \frac{a}{u}\right) du = \frac{1}{b} (u - a \cdot \log_e u) + c$$

Hence, on substituting back for u,

$$\int \frac{bx\,dx}{a + bx} = \frac{1}{b} \{(a + bx) - a \cdot \log_e(a + bx)\} + c \qquad (10.33)$$

Finally, substituting 10.33 into 10.32, we have

$$\int \{\log_e(a + bx)\}dx = x \cdot \log_e(a + bx) - \frac{1}{b} \{(a + bx) - a \cdot \log_e(a + bx)\} + c$$

$$(10.34)$$

GROWTH FUNCTIONS

One of the areas of biology to which mathematical analysis has usefully been employed is in the analysis of growth of cells, whole organisms or populations. We have already used an example applying calculus to a biological problem, namely the growth of a microbial culture. In that example, we formulated, in mathematical terms, our initial knowledge of the way in which the rate of increase of cell number should change in relation to the number of cells actually present. We arrived at an equation summarizing this knowledge, 6.12. Because this equation contains a differential coefficient, it is called a differential equation, and it relates increase in cell number, n, to cell number present at time t:

$$\frac{dn}{dt} = f(n) \qquad (10.35)$$

It has already been stated that from this differential equation we can obtain another equation giving a relationship between cell number and time, i.e.

$$n = F(t) \qquad (10.36)$$

Assuming that the microbial colony completely obeyed the assumptions we made, and therefore also the equations derived for it, the number of cells present at any selected time t is given by equation 10.36; and the curve of 10.36 would appear as in Fig. 6.5(b). An equation of the form of 10.36 is called a **growth function**, and its curve may be known as a **growth curve**; but the latter term is often applied to a plot of actual observations as shown in Fig. 6.5(a), especially if adjacent points are joined by straight lines.

Now although the differential equation 6.12 is simple in form, that is $f(n)$ in 10.35 is simple, the resulting $F(t)$, 10.36, is not an equally elementary function. So first we shall consider an even simpler differential equation giving rise to a growth function. Since quantities other than cell number may be used for describing growth, we shall use y instead of n in what follows.

The exponential function

We have remarked that, when the number of cells in a microbial culture is small relative to the quantity of medium in which the colony finds itself, the rate of growth at any instant of time t would be proportional to the size of the colony (number of organisms present), y, at that time (page 105). This could be put into mathematical notation as

$$\frac{dy}{dt} = by \qquad (10.37)$$

On page 184, it was stated that when the differential equation 10.37 is solved to give y as a function of time, the result is the exponential function

$$y = ae^{bt} \qquad (10.38)$$

where $a = e^c$, c being the constant of integration. Since 10.38 is an exponential function, any biological material, from a cell to a whole population of multicellular organisms, which is growing in such a manner that it can be described by equation 10.38 is said to be **growing exponentially**.

For any particular set of growth data, how can we establish whether it conforms to the exponential function or not? To do this rigorously involves concepts in statistics, and is therefore outside the scope of this book; but one can get some idea from a graphical examination of the data. However, it would not be very illuminating to plot the growth measurements, y, obtained at a number of times, t, on a t–y graph and try to see whether, assuming there was no sample to sample variability (page 104), the data lay on an exponential curve. Even if there were no sampling variability, it would be impossible to assert that the data followed the curve of an exponential function rather than that of one of many other types of function. The only type of curve that it is possible to distinguish with any degree of certainty is a straight line. In the case of the exponential function, we know that the relationship between $\log_e y$ and t is linear, 10.16; hence, if the logarithms of our growth measurements are plotted against time, as in Fig. 10.1(c), and if the resulting points look as though they would lie on a straight line if sampling variability were absent, then it can be inferred that exponential growth underlies the data produced.

Now although growth cannot be exponential over long periods, such a pattern is found in many biological situations over short periods of time. Thus growth is exponential, or approximately so, in the early stages of growth of a microbial culture. In higher plants, growth of the seedling for a while after it has become independent of seed reserves is often said to be exponential. As far as animals are concerned, it seems that exponential growth only occurs in the embryonic stage.

Absolute and relative growth rates

It may be asked just what form does a growth function and its curve take if it is not exponential? To answer this, we need to return to the growth rate form of the exponential function, equation 10.37. Now in biology, the growth rate as defined by dy/dt is often not a very useful concept, and another form of growth rate is usually more meaningful. In order to distinguish between the two types of growth rate, they are given different names; the growth rate shown in equation 10.37 as dy/dt is called the **absolute growth rate**. So if growth is exponential, this means that at any instant of time the absolute growth rate is proportional to size already attained. The last sentence indicates why the absolute growth rate is not very helpful to the growth analyst: young sunflower plants have a higher absolute growth rate than plants of chickweed at the same stage of development, and most of the difference in absolute growth rate between the plants of these two species is due to the size difference, rather than to any difference in the constants of proportionality, b.

Now if both sides of 10.37 are divided by y (the size), we have

$$\frac{1}{y} \cdot \frac{dy}{dt} = b \tag{10.39}$$

which provides a new definition of rate of growth. The quantity on the left-hand side of 10.39 is called the **relative growth rate**, or sometimes the **specific growth rate**; it is the absolute growth rate divided by size, or the absolute growth rate per unit size. Viewed in this way, it is evident that the *relative growth rate is a measure of the efficiency of the material in producing new material.*

When growth is exponential, we see from 10.39 that this efficiency is constant; in particular, it does not change with time. If relative growth rate is increasing with time, we could describe this as supra-exponential growth. This type of growth only occurs over very limited periods in well-defined circumstances, e.g. when a new bacterial culture is changing from its lag phase to its exponential phase, or in a very young seedling when photosynthesis is rapidly increasing and overcoming the loss in dry weight due to respiration. It is always important when considering growth to state what aspect of growth is being studied. In the above examples, it is cell number in the microbial culture, while growth in higher plants is normally measured in terms of dry weight.

After possibly going through a period of exponential growth, an organism continues growing but with a *declining relative growth rate*. For many, or perhaps most, organisms all their growth is made in this way, and so to describe such growth mathematically we need to look for a function of the form $y = f(t)$ in which $(1/y) \cdot (dy/dt)$ is continuously decreasing.

Before leaving this section, and examining two functions with the above property, it should be noted that relative growth rate is often given its own symbol, R, and so

$$R = \frac{1}{y} \cdot \frac{dy}{dt} \qquad (10.40)$$

When the exponential function is used to describe growth, i.e. equation 10.38, we see from equations 10.39 and 10.40 that $R = b$; so the exponential growth function may be written as

$$y = ae^{Rt} \qquad (10.41)$$

which emphasises the fact that when growth is exponential, relative growth rate is constant.

The logistic function

Reverting yet again to our old friend the microbial culture, we saw that the differential equation 6.12 described the absolute growth rate of the culture from (in theory) no individuals to (again in theory) the maximum number of individual cells the culture could contain. Thus the equation is an adequate overall description of events throughout the growing life of the colony. If we divide both sides of 6.12 by y (using the more general

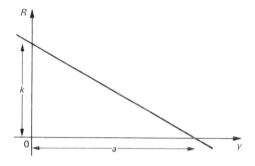

Fig. 10.3   The straight line $R = k - ky/a$.

symbol y instead of n), we have

$$\frac{1}{y} \cdot \frac{dy}{dt} = k \left(1 - \frac{y}{a} \right)$$

i.e.

$$R = k - \frac{k}{a} y \qquad (10.42)$$

Relationship 10.42 shows the relative growth rate to be a linear function of size, with gradient $-k/a$ and intercept k (Fig. 10.3). Evidently, relative growth rate is nowhere constant, but declines continuously (in fact, linearly) with increasing size. The growth function obtained by solving the differential equation (we shall do this in Chapter 13, pages 246–7) is

$$y = \frac{a}{1 + be^{-kt}} \qquad (10.43)$$

where $b = e^c$, c being the constant of integration.

We already know that the maximum size is a, but this fact may also be deduced directly from equation 10.43. As $t \to \infty$, $e^{kt} \to \infty$; hence $e^{-kt} \to 0$. Thus $be^{-kt} \to 0$, and so $(1 + be^{-kt}) \to 1$. So we see that as $t \to \infty$, $y \to a$; the curve is asyn.ptotic to the maximum size a. The intercept on the y-axis is found by putting $t = 0$ in 10.43; we can call it y_0. When $t = 0$, $e^{-kt} = 1$, and so $be^{-kt} = b$. Thus

$$y_0 = \frac{a}{1 + b} \qquad (10.44)$$

The function 10.43 has various names, the two commonest being the **logistic** and the **autocatalytic**. The curve of the function is shown in Fig. 10.4(a), and it can be seen that the gradient (the absolute growth rate) is positive throughout. The gradient increases in value near the beginning

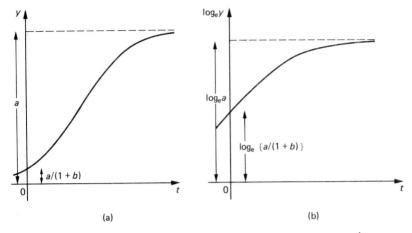

Fig. 10.4   The logistic function: (a) curve of the function $y = a/(1 + be^{-kt})$, (b) curve of the function $\log_e y = \log_e a - \log_e(1 + be^{-kt})$.

and decreases towards the end of the growth period so there is a point of inflexion: this occurs at $y = a/2$, that is, when half the total growth has been made.

Logarithmic form
    For a number of reasons, it is most convenient to deal with growth data in logarithmic form. One of these reasons is that a test for exponential growth can be made (page 196) when the data are in this form; another reason is that growth functions have to be fitted to data in logarithmic form. The equation of the logarithmic form of the logistic function can be obtained by taking natural logarithms of both sides of 10.43 to give:

$$\log_e y = \log_e a - \log_e(1 + be^{-kt}). \tag{10.45}$$

The curve for this is shown in Fig. 10.4(b). There is no point of inflexion, and the gradient, though always positive, declines throughout. The intercept is now $\log_e\{a/(1 + b)\}$, which could be negative, while the asymptote is, of course, $\log_e a$.

Linear form
    A linear form of the logistic function can be obtained by first rearranging 10.43 to give

$$\frac{a}{y} = 1 + be^{-kt}$$

and so

$$\frac{a}{y} - 1 = be^{-kt}$$

Taking natural logarithms of both sides gives

$$\log_e \left(\frac{a}{y} - 1 \right) = \log_e b - kt \tag{10.46}$$

Thus $\log_e(a/y - 1)$ is a linear function of t, with a gradient of $-k$ and an intercept of $\log_e b$.

Relationship 10.46 could be used to test whether a set of data reasonably conforms to a logistic function, provided an estimate of a can be supplied. This could be done if the data in hand provided observations near to or at the maximum size. Then for each of the size observations, y, at time t, calculate $\log_e(a/y - 1)$, and plot the results against t. If the points on the graph approximate to a straight line, then the growth data approximate to a logistic function.

At a first reading, skip to Chapter 11 if you are interested in plant growth analysis; otherwise proceed to Chapter 13.

The Gompertz function

In 1940, P. B. Medawar published a paper in which he deduced theoretically that the growth of the embryo chicken heart should follow a particular kind of growth function. The function concerned is known as the **Gompertz**, which was formulated as long ago as 1825 by a German mathematician of that name, in connexion with actuarial studies. The equation is

$$y = ae^{-be^{-kt}} \tag{10.47}$$

and the logarithmic form is

$$\log_e y = \log_e a - be^{-kt} \tag{10.48}$$

The curves of functions 10.47 and 10.48 are shown in Fig. 10.5, (a) and (b), respectively.

Differentiating 10.47 with respect to time, using the function of a function rule, gives

$$\frac{dy}{dt} = akbe^{-kt}e^{-be^{-kt}}$$

and dividing both sides by y gives the relative growth rate as a function of

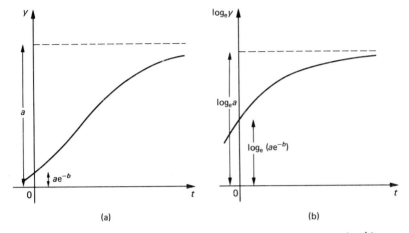

Fig. 10.5   The Gompertz function: (a) curve of the function $y = ae^{-be^{-kt}}$, (b) curve of the function $\log_e y = \log_e a - be^{-kt}$.

time:

$$R = kbe^{-kt} \tag{10.49}$$

Taking logarithms of both sides of (10.49) gives

$$\log_e R = \log_e(kb) - kt \tag{10.50}$$

Thus, the logarithm of relative growth rate is a linear function of time, with gradient $-k$ and intercept $\log_e(kb)$.

Another linear relationship involving relative growth rate may be derived as follows. From 10.49, we have

$$be^{-kt} = \frac{R}{k}$$

and from 10.48

$$be^{-kt} = \log_e a - \log_e y$$

Since the left-hand sides of the above two equations are equal, we may equate the right-hand sides:

$$\frac{R}{k} = \log_e a - \log_e y$$

i.e.

$$R = k \cdot \log_e a - k \cdot \log_e y \tag{10.51}$$

Equation 10.51 shows that the relative growth rate is a linear function of the logarithm of size, with gradient $-k$ and intercept $k \cdot \log_e a$. Thus the Gompertz function has two linear relationships involving relative growth rate, one with time, and the other with the logarithm of size.

Finally, we may obtain a linear form of the growth function itself from equation 10.48:

$$\log_e\left(\frac{a}{y}\right) = \log_e b - kt \qquad (10.52)$$

Medawar's growth model

When Medawar deduced theoretically that the growth of the embryo chicken heart should follow a Gompertz function, he used the concept of 'growth energy'. He argued that this growth energy should be assessable through the application of a specific growth inhibitor; the more inhibitor required to stop growth, the higher the growth energy.

The experimental basis consisted of the application of just sufficient inhibitor to make the growth of the tissue cease, at a series of different ages. Then, assuming that growth energy was directly proportional to the amount of inhibitor required, it was found that when the logarithm of growth energy, $\log_e G$, was plotted against time, t, the trend of the points was linear with a negative gradient

$$\log_e G = \log_e A - Kt \qquad (10.53)$$

where A and K are constants. Medawar also found a linear relationship to exist between growth energy and the logarithm of the mass of the tissue, $\log_e W$,

$$G = B - K \cdot \log_e W \qquad (10.54)$$

where B and K are constants. If the forms of 10.53 and 10.54 are compared with 10.50 and 10.51, they are found to be identical. Thus, growth energy is a measure of relative growth rate, and the growth of the tissue must conform to the Gompertz function.

EXERCISES 10

1. Without using mathematical tables, evaluate:

(a) $e^{0.3}$ (b) $e^{-0.2}$ correct to 4 decimal places.

2. Plants of birch (*Betula pubescens*) and sycamore (*Acer pseudoplatanus*) were grown from seed, and harvests were taken at two-weekly intervals during the first season. Growth was found to be exponential for 18 weeks, and the equations fitted to the data were:

for birch:

$$W_B = 0.0441e^{0.3041t}$$

for sycamore:

$$W_S = 0.3837e^{0.2243t}$$

where W was measured in g, and t in weeks from the first harvest.
(a) Write down the relative growth rate of each species.
(b) Write down the weight of each species at the first harvest.
(c) Calculate the weight of each species at the end of the exponential phase.
(d) If the above growth function had been applicable over a sufficiently long period, find the time when the weight of each of the two species would be identical.
3. In a sample of NaH_2PO_4 containing a known amount of ^{32}P it was found that 95.24% of the ^{32}P remained after 24 hours. What is the half-life of ^{32}P?
4. In an experiment to determine the effects of the plant hormone abscisic acid on the growth of isolated lengths of oat coleoptiles, it was found that the following equation empirically described the data:

$$y = \log_e(354\ 123x^{-0.4721})$$

where x is the hormone concentration in parts per million, and y is the final length of the coleoptiles in mm 4 hours after treatment with the hormone. What is the expected final length of the coleoptiles after 4 hours when treated with:

(a) 0.1 ppm, and   (b) 5 ppm of abscisic acid?

5. Differentiate the following functions with respect to x:

(a) $e^{(a+bx)}$ (b) $e^{(1-2x+x^2)}$ (c) $1 + be^{-kx}$ (d) $\log_e(a + bx + cx^2)$

6. Find   (a) $\int dx/(1+x)$ (b) $\int dx/(3x-7)$ (c) $\int(1 + be^{-kt})dt$
7. A colony of *Drosophila* flies was bred on an artificial medium, and the number of individuals, n, at time t (days) was found to conform to the logistic function:

$$n = \frac{230}{1 + 6.8888e^{-0.1702t}}$$

$n = \dfrac{a}{1 + be^{-kt}}$

(a) Write down the theoretical maximum number of individuals in the colony.
(b) How many flies were originally present in the colony?
(c) How many flies would be expected on day 10?

(d) Re-arrange the equation to give t in terms of n, instead of n in terms of t. Then calculate the time by which the maximum number is effectively reached ($n = 229.5$).

8*. Integrate by parts:

(a) $\int xe^x dx$ (b) $\int x^2 e^x dx$ (c) $\int x(\log_e x)dx$ (d) $\int \{(\log_e x)/x^2\}dx$

9*. Evaluate   (a) $\int_0^1 xe^x dx$ (b) $\int_0^\infty xe^{-x}dx$

II

Plant Growth Analysis:
A Biological Application of Calculus

GROWTH OF MICRO-ORGANISMS, PLANTS, AND ANIMALS

We have already considered at some length various aspects of the mathematical analysis of the growth of a microbial culture. First, experimental data were plotted on a graph of cell number against time (Fig. 6.5(a)), and then we assumed that if sampling variability were absent the growth of the culture should follow some smooth curve, e.g. as shown in Fig. 6.5(b). Then, instead of merely trying a number of different mathematical functions to see if we could find a curve which would fit the data well, we tried to build a working mathematical model of the growth of the culture based on known biological facts and principles. Such considerations led us to the differential equation 6.12, which related rate of increase in cell number at any instant of time to cell number actually present at that time. Solving this differential equation gave another function relating cell number to time; this is the function we were seeking to fit to our graph of the experimental data. In the previous chapter, we saw that the solution of the differential equation 6.12 gave rise to the logistic function.

When it comes to higher organisms, with multicellular bodies, things are not so simple. One can, of course, regard the multicellular organism as a colony of individual cells, but the main difference between a colony of unicellular organisms and a single multicellular organism is that in the latter there is a considerable interaction between the constituent cells. This results in a growth promotion of some cells, and an inhibition in the growth of others. There is 'division of labour' between cells, and so they differentiate into various types of tissue. Hence it might appear that a simplified model of growth on the lines developed earlier in the book will not suffice for higher organisms.

Nevertheless, growth data obtained for such organisms (i.e. measurements of a certain aspect of size, usually fresh weight in animals and dry weight in plants) often show curves which are similar in form to that developed for a microbial colony. The main difference will be in the position of the point of inflexion; in the logistic function, this point of maximum absolute growth rate is mid-way between zero and maximum size, while in the growth curves of higher organisms the point of inflexion is usually elsewhere.

Among the higher organisms, there is a fundamental difference between animals and plants in respect of growth. The animal body normally shows a definite period of growth, the immature or juvenile phase, and when the animal is mature growth usually ceases. During the growth phase, the different parts of the animal body (head, limbs, etc.) are normally growing in a synchronized manner; in fact, it is found that the relative growth rates of the different parts of the animal's body bear a constant ratio to one another. Because of the cessation of size increase at maturity, growth is said to be **determinate**.

The situation is different in plants. Taking flowering plants, an annual species is a familiar and simple example. It is still true that we can distinguish a juvenile phase when vegetative growth is being made, and an adult phase when growth of the reproductive organs is occurring. The latter marks an important difference between plants and animals, in that the reproductive parts of a plant form a much greater proportion of the overall size than do the reproductive organs of animals. It is also true that the relative growth rates of the major parts of the plant — foliage, stem, roots — may bear a constant relationship to one another for prolonged periods of time, but it must be remembered that each of these main organ systems does not have a homogeneous structure. Thus the foliage is made up of leaves all of different ages, the stem is likewise composed of many internodes, and even the roots can be subdivided in a similar manner.

Many plants continue to grow vegetatively even during the reproductive phase; for example, new leaves continue to be initiated at the shoot apex if the inflorescence is axillary. Now although the growth of an *individual* leaf is determinate, that is it grows to a certain maximum size and then stops, the growth of the foliage system is **indeterminate** since the ability of the plant to produce new leaves is not usually definitely circumscribed. In fact, since each of the main organ systems on a plant is composed of subunits, the growth of the whole plant is indeterminate. Hence, if the growth of an annual plant is followed from germination to senescence, although the growth data may tend to approach something resembling a sigmoid curve (when the growth attribute is plotted against time), the whole situation is much more nebulous than in the case of animals or micro-organisms. Even if a mathematical function could be found to fit growth data obtained from whole plants, it would not be very meaningful for two reasons. Firstly, the function would tend to be so complex that a useful biological interpretation of it would be very difficult, and secondly, the indeterminacy of growth would lead to a model which was not simple enough for the constants in it to have a useful biological meaning. Hence for plants, the analysis of growth is not attempted in terms of a fitted mathematical function, but methods have been developed, based directly on the actual growth data themselves, which enable useful biological infor-mation to be obtained. The scheme, which is known simply as '**plant growth analysis**', is described here as it affords a good illustration of an application of the calculus to a problem in biology.

PLANT GROWTH ANALYSIS: PRINCIPLES

Although the methods of plant growth analysis do not rely on any mathematical function fitted to the growth data, it is convenient in developing the ideas to imagine a group of annual plants whose growth happens to conform very closely to that of a logistic function. This means that the growth curve of the plants will be as shown in Fig. 11.1(a). There are many ways of defining and measuring plant growth, and the method selected will depend on the purpose of the study. Usually, growth studies in plants are enquiries into their productivity, measured in terms of their increase in non-aqueous material; so the attribute measured is dry mass, or as it is more commonly (but erroneously) called, dry weight. As dry weight is a destructive measurement, it is not possible to make a series of measure-ments on one plant; a group of similar plants must be used, and the assumption is made that the patterns of growth shown by the individual plants in the group are very similar.

Dry weight is normally given the symbol of W, while time is denoted by t as usual. With this notation, absolute growth rate is dW/dt, and relative growth rate is $(1/W)(dW/dt) = R$. When the $W = f(t)$ relationship is logistic, then the curves of absolute and relative growth rates with respect to time are as shown in Fig. 11.1(b) & (c), respectively. As remarked in the previous chapter (page 196), absolute growth rate is not a useful concept, and in fact does not appear in the methodology of plant growth analysis; thus we need concern ourselves no longer with absolute growth rate. The curve of relative growth rate declines throughout growth in a 'reverse sigmoid' manner: the change in relative growth rate as the plant grows is an important factor in plant growth analysis, and we shall return to consider the measurement of relative growth rate later on. Meanwhile,

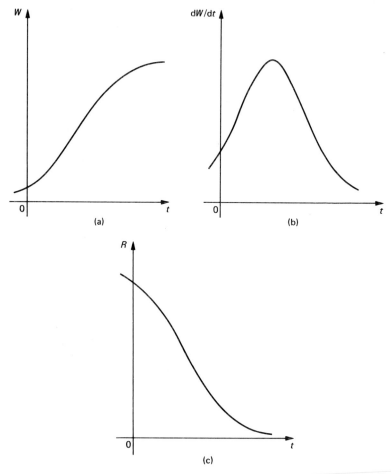

Fig. 11.1 The logistic as a growth function: (a) W as a function of time, (b) absolute growth rate dW/dt as a function of time, (c) relative growth rate $(1/W)(dW/dt)$ as a function of time.

there are other quantities involved in plant growth analysis, which must now be considered.

The partitioning of relative growth rate

On page 197 it was stated that the relative growth rate of an organism is a measure of the efficiency of the material comprising the organism in the production of new material, and evidently this measure is an average over the whole organism. Now, as we have already pointed out, a higher plant is far from homogeneous, and so a measure of the average efficiency of dry matter production is limited in the information it can convey.

By far the largest part of the dry matter of a plant is in the form of organic compounds, whose primary constituent, carbon, was originally assimilated into the plant through the leaves in the form of carbon dioxide. Only a small fraction of the plant's dry mass consists of mineral elements assimilated from the soil via the roots. Hence if, as a first approximation, the mineral content is disregarded, the leaves are the sole assimilatory organs by means of which material external to the plant becomes incorporated into the plant body. Therefore, any measure of the overall efficiency of the plant depends partly on the leaves; and so the actual value of the relative growth rate at an instant of time will depend on the efficiency of the leaves in assimilation, and the amount of foliage present, at that time. If the latter is measured as the proportion of leaf dry weight to dry weight of the whole plant, we can write

$$R = E \cdot \frac{L}{W} \qquad (11.1)$$

where L is the dry weight of the foliage of the plant, and W is the dry weight of the whole plant (including the leaves). The ratio L/W evidently represents the amount of foliage relative to the whole plant, and so E is a measure of the efficiency of the leaves in the production of new dry matter. Expanding the left-hand side of 11.1, we have

$$\frac{1}{W} \cdot \frac{dW}{dt} = E \cdot \frac{L}{W}$$

and cross-multiplication yields

$$E = \frac{1}{L} \cdot \frac{dW}{dt} \qquad (11.2)$$

The right-hand side of this expression is very similar to relative growth rate; it is the absolute growth rate of the whole plant divided by leaf dry weight, or the absolute growth rate of the plant per unit leaf dry weight, and can be interpreted as the efficiency of the leaf material in the production of new dry matter. The quantity E is called the **net assimilation rate**, and is a measure of net photosynthetic rate. The ratio L/W in relationship 11.1 is termed the **leaf-weight ratio**.

The partitioning of net assimilation rate

When measuring the photosynthetic efficiency of a plant, the plant physiologist usually expresses the rate per unit of leaf area. This attribute of the leaf is usually more relevant in photosynthetic studies than is leaf dry weight. It is the area of the laminae which determines how much radiation is intercepted, and the number of stomata, through which carbon dioxide diffuses, is more dependent on leaf area than on leaf weight. As we have defined it in the previous paragraph, it would appear that net assimilation rate must be assessed per unit of leaf dry weight. Yet this is not so. Suppose that net assimilation rate is defined per unit leaf area, then 11.2 becomes

$$E_A = \frac{1}{L_A} \cdot \frac{dW}{dt}$$

The suffix, A, indicates that we are using leaf area, L_A, and are measuring net assimilation rate on a leaf area basis, E_A. Net assimilation rate on a leaf weight basis 11.2 could be written as

$$E_W = \frac{1}{L_W} \cdot \frac{dW}{dt}$$

and it is E_W which enters into the primary partitioning of relative growth rate, equation 11.1. Now

$$\left. \begin{array}{l} \dfrac{1}{L_W} \cdot \dfrac{dW}{dt} = \dfrac{1}{L_A} \cdot \dfrac{dW}{dt} \times \dfrac{L_A}{L_W} \\[3mm] E_W = E_A \times \dfrac{L_A}{L_W} \end{array} \right\} \tag{11.3}$$

i.e.

Substituting for E_W (i.e. E) in 11.1, we finally have

$$R = E_A \times \frac{L_A}{L_W} \times \frac{L_W}{W} \tag{11.4}$$

where L_W has been written instead of L in the last term on the right-hand side.

Thus relative growth rate, which measures the efficiency of the plant as a whole in the production of dry matter, has been resolved into three components, each with a distinct physiological or morphological meaning. Net assimilation rate, E_A, is a measure of net photosynthetic rate, as already mentioned. Very broadly, photosynthesis may be regarded as a process with an input and an output; the input is carbon dioxide and water, and the output is oxygen plus organic compounds. Net assimilation rate is a unique method of assessing photosynthetic rate in that it does so by measuring the rate of production of organic matter. All other methods of measuring the photosynthetic rate involve the rate of input of carbon dioxide or the rate of output of oxygen. Methods based on gaseous exchange measure the photosynthetic rate over a relatively short time period, whereas net assimilation rate is, as we shall see later, a long-term measure.

The second term on the right-hand side of 11.4 is the ratio of leaf area to dry weight, and is called the **specific leaf area**. To some extent it is a measure of leaf thickness, but there is more to it than that. Consider a single leaf: as a primordium it consists of meristematic cells without vacuoles and intercellular spaces; there is plenty of cellular material there but very little area, and hence the specific leaf area is low. As the leaf expands, cell division slows down, and cell extension, with its accompanying vacuolation and formation of intercellular spaces, becomes predominant. Area increases faster than dry mass, and so the specific leaf area rises. After maximal area has been reached, accumulation of dry matter still persists in the form of deposition of extra cell wall material, and the specific leaf area declines. Specific leaf area may thus be regarded as a measure of leaf density.

The final term, L_W/W, is the leaf-weight ratio, and is quite simply the ratio of leaf dry weight to total plant dry weight. Hence, this ratio lies in the range $0 \leqslant L_W/W < 1$. In the very young seedling of species whose seeds contain endosperm, the leaf-weight ratio is small at first, while the seedling is dependent on endosperm reserves. As the plant becomes nutritionally independent, the leaf-weight ratio increases rapidly and may reach a value of about 0.8 in herbaceous species, but rather lower in woody types. In many herbaceous plants the leaf-weight ratio stays constant during vegative growth, especially if the form of growth is a rosette. When the stem grows, the leaf-weight ratio falls; again, the decline in leaf-weight ratio is particularly marked in woody species. During the reproductive phase, the decline in leaf-weight ratio is accelerated.

MEASUREMENT OF THE QUANTITIES INVOLVED IN GROWTH ANALYSIS

Specific leaf area and leaf-weight ratio

These ratios present no problem. The primary data for a growth analysis consist of determinations of the dry weights of the leaves and the rest of the plant, together with an estimate of total leaf area, for each member of a batch of plants taken at a particular point in time. The batch of plants taken at a particular time is known as a harvest; several harvests are taken, usually at regular intervals. At each of the harvests, specific leaf area and leaf-weight ratio are simply evaluated from the data.

Relative growth rate

This parameter has been defined as $(1/W)(dW/dt)$; it contains a differential coefficient, which implies that it is defined at an instant of time. It is not possible to measure an instantaneous rate of change, and even if it were, the result would not be meaningful. To see this, remember that plant growth has been defined in terms of change in dry weight, which is a direct consequence of photosynthesis and respiration. Now the primary environmental factor affecting photosynthetic rate, and hence the rate of dry weight increase, is light intensity. Photosynthetic rate in this context is net, i.e. photosynthetic rate minus respiration rate, and evidently it will be negative at night. Therefore, relative growth rate will also be negative at night. Moreover, on partly cloudy days the relative growth rate will be frequently changing. Even without short term changes in light intensity during daylight hours due to cloud, a curve of relative growth rate against time would look like that in Fig. 11.2 over a succession of days and nights. Hence at any instant of time, a value for relative growth rate may be almost unique.

However, it is readily apparent that over the three day period shown in Fig. 11.2, a *mean* relative growth rate, $\bar{R}$, is a valid concept (page 150), and it is shown as a horizontal line on the graph. So if we imagine that a harvest is taken at sunrise on the first day (at $t = 0$), and another harvest is taken at sunrise on the fourth day, the interval between the harvests is 72 hours. How can we calculate the mean relative growth rate over the three day period?

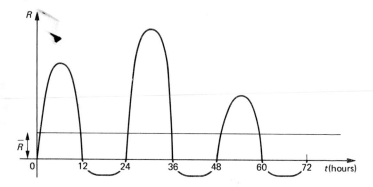

Fig. 11.2   A typical curve of relative growth rate against time for a plant over a three-day period. The daylength is 12 hours and it is assumed that there are no short-term fluctuations in light intensity. The mean relative growth rate, $\bar{R}$, for the 72-hour period is also shown.

The principle of the method for answering this question was given in Chapter 8 (page 152). The curve shown in Fig. 11.2 must be the curve of some function relating relative growth rate to time, i.e. a function of the form

$$R = f(t) \tag{11.5}$$

We may never know what $f(t)$ is; indeed it is highly unlikely that we could ever find a function which would generate the curve in Fig. 11.2 exactly. However, for the situation shown, and measuring time in days, we have from equation 8.15 that

$$\bar{R} = \frac{1}{3 - 0} \int_0^3 f(t)dt$$

More generally, we consider two harvests, one at time t_1 and the other at time t_2; when the mean relative growth rate between these times is given by

$$\bar{R} = \frac{1}{t_2 - t_1} \int_{t_1}^{t_2} f(t)dt \tag{11.6}$$

Now although the form of $f(t)$ is unknown, it can be replaced by R, see 11.5, and in turn, R can be replaced by $(1/W)(dW/dt)$. Thus 11.6 becomes

$$\bar{R} = \frac{1}{t_2 - t_1} \int_{t_1}^{t_2} \frac{1}{W} \cdot \frac{dW}{dt} \cdot dt \tag{11.7}$$

Now in 11.7 to the right of the integral sign, the dt's cancel leaving dW/W. So the variable of integration is now dW instead of dt, and because of this the limits of integration cannot remain as t_1 and t_2. Let plant weight at the first harvest be W_1 and at the second harvest W_2, so 11.7 becomes

$$\bar{R} = \frac{1}{t_2 - t_1} \int_{W_1}^{W_2} \frac{dW}{W} \tag{11.8}$$

Note that the $1/(t_2 - t_1)$ outside the integral sign remains in terms of t because it is a constant. Integrating:

$$\bar{R} = \frac{1}{t_2 - t_1} \left[\log_e W \right]_{W_1}^{W_2}$$

(see standard integral 10.24), and so

$$\bar{R} = \frac{\log_e W_2 - \log_e W_1}{t_2 - t_1} \tag{11.9}$$

Since the original function relating R and t, 11.5, is completely unknown, and no assumptions have been made about it in the above derivation, equation 11.9 will always give the correct mean relative growth rate between two harvests regardless of the way the plants are growing during the interval.

In this chapter, two graphs depicting curves of R against time have been presented (Fig. 11.1(c) and Fig. 11.2) which look very different from one another. The true situation for growing plants is shown in Fig. 11.2, while the curve shown in Fig. 11.1(c) has been derived from a relatively simple mathematical function relating increase in plant weight to time. This latter function is continually increasing, and so can never yield a negative relative growth rate which undoubtedly occurs at night. So the curve shown in Fig. 11.1(c) should be regarded more as a curve of mean relative growth rate against time: it is a smooth approximation to an $\bar{R}-t$ curve rather than an $R-t$ curve.

Net assimilation rate

All the remarks made about relative growth rate in the first two paragraphs of the previous section apply equally to net assimilation rate, and so in practice we calculate the mean net assimilation rate between two harvests. Suppose that net assimilation rate is related to time by some function

$$E = \phi(t) \tag{11.10}$$

Then between two harvests at t_1 and t_2 the mean net assimilation rate is given by

$$\bar{E} = \frac{1}{t_2 - t_1} \int_{t_1}^{t_2} \phi(t) dt$$

Now $\phi(t) = (1/L)(dW/dt)$, where L is leaf area in this context, and so

$$\bar{E} = \frac{1}{t_2 - t_1} \int_{t_1}^{t_2} \frac{1}{L} \cdot \frac{dW}{dt} \cdot dt \tag{11.11}$$

As in the similar expression for relative growth rate 11.7, the dt's cancel leaving dW as the variable of integration, and so the limits t_1 and t_2 are replaced by W_1 and W_2:

$$\bar{E} = \frac{1}{t_2 - t_1} \int_{W_1}^{W_2} \frac{dW}{L} \tag{11.12}$$

The integral in 11.12 cannot be evaluated, however, unless the relationship between W and L is known or assumed.

We have not discussed relationships between plant parts in this book (except for a brief comment on page 206), and this is quite a large subject in itself. Usually, however, during the vegetative phase of growth it is safe to assume that there is a linear relationship between total plant weight and leaf area, i.e.

$$W = a + bL \tag{11.13}$$

and the error involved if the relationship departs appreciably from linearity is minimized if the harvest is such that leaf area no more than doubles from one harvest to another. Differentiating 11.13 gives

$$\frac{dW}{dL} = b$$

or

$$dW = bdL \tag{11.14}$$

On using 11.14 to substitute for dW in 11.12, we again have a new variable of integration, dL, and so the limits of integration in 11.12 must be changed once more. Assume that at t_1, when total plant weight is W_1, the leaf area is L_1; similarly at t_2, when total plant weight is W_2, the leaf area is L_2. Putting all these changes into 11.12, it becomes

$$\bar{E} = \frac{1}{t_2 - t_1} \int_{L_1}^{L_2} \frac{bdL}{L}$$

or

$$\bar{E} = \frac{b}{t_2 - t_1} \int_{L_1}^{L_2} \frac{dL}{L} \tag{11.15}$$

Integrating,

$$\bar{E} = \frac{b}{t_2 - t_1} \left[\log_e L \right]_{L_1}^{L_2}$$

and so

$$\bar{E} = \frac{b(\log_e L_2 - \log_e L_1)}{t_2 - t_1} \tag{11.16}$$

Now although to get this far we have had to assume a linear relationship between W and L, we do not wish to assign actual values to the constants in this relationship; fortunately this is not necessary. At the harvest at t_1 we have, from equation 11.13 and the paragraph beneath 11.14,

$$W_1 = a + bL_1 \tag{11.17}$$

and at t_2

$$W_2 = a + bL_2 \qquad (11.18)$$

Subtracting 11.17 from 11.18, we have

$$W_2 - W_1 = (a - a) + (bL_2 - bL_1)$$

i.e.

$$W_2 - W_1 = b(L_2 - L_1)$$

which gives

$$b = \frac{W_2 - W_1}{L_2 - L_1} \qquad (11.19)$$

Substituting for b in 11.16 gives

$$\bar{E} = \frac{(W_2 - W_1)(\log_e L_2 - \log_e L_1)}{(L_2 - L_1)(t_2 - t_1)} \qquad (11.20)$$

Equation 11.20 gives the correct value for the mean net assimilation rate between the two harvests regardless of how the whole plant and leaf area are growing with respect to time, but subject to the constraint that there is a linear relationship between whole plant dry weight and leaf area.

Thus all the quantities involved in plant growth analysis can be calculated from primary data obtained at each of a number of harvests — a minimum of two is required. The two ratios — leaf-weight ratio and specific leaf area — are calculated from leaf weight, whole plant weight, and leaf area at each harvest. The mean relative growth rate between an adjacent pair of harvests is computed from the whole plant weights at each of the two harvests, and mean net assimilation rate between a pair of harvests by using the leaf areas and total plant weights at each of the two harvests.

EXERCISE 11

The following data are the leaf areas, L_A, leaf dry weights, L_W, and total plant dry weights, W, of two series of plants of *Browallia speciosa* at a series of times, t. One series was grown in full daylight, and the other was grown under a screen admitting about 40% of full daylight. Each figure is the mean of 4 plants.

	UNSHADED			SHADED		
t (days)	L_{A_2} (cm²)	L_W (g)	W (g)	L_{A_2} (cm²)	L_W (g)	W (g)
0	0.3000	0.0009	0.0012	0.5093	0.0007	0.0011
3	0.5000	0.0014	0.0020	0.7596	0.0011	0.0015
7	1.4787	0.0037	0.0048	0.8293	0.0011	0.0015
10	1.3835	0.0031	0.0042	1.1960	0.0019	0.0026
14	2.4177	0.0070	0.0096	2.3262	0.0034	0.0048
17	3.4185	0.0104	0.0149	2.5225	0.0044	0.0064
21	7.9747	0.0205	0.0301	2.9430	0.0043	0.0060
24	5.8193	0.0157	0.0234	5.0663	0.0074	0.0106
28	10.4533	0.0315	0.0460	6.0600	0.0100	0.0139

Calculate, for each series of plants, L_A/L_W and L_W/W at each harvest, and $\bar{R}$ and $\bar{E}$ for each harvest interval. Draw graphs of L_A/L_W, L_W/W, $\bar{R}$, and $\bar{E}$ against t to show the effect of shading on these quantities. Try to interpret the results biologically.

12

Trigonometric Functions

TRIGONOMETRIC FUNCTIONS AND THEIR BIOLOGICAL USES

The trigonometric functions of x are $\sin x$, $\cos x$, $\tan x$, $\operatorname{cosec} x$, $\sec x$, $\cot x$. In words, these functions are: **sine, cosine, tangent, cosecant, secant, cotangent,** of x. You will have already met at least the first three of these in relation to an acute angle, θ, of a right-angled triangle. In Fig. 12.1 is shown a right-angled triangle ABC with the right-angle at C. In relation to the acute angle A, the three sides of the triangle are named thus:

Some People Have side BC is called 'opposite' *Perpendicular* $\sin = \frac{P}{H}$
Curly Black Hair side AC is called 'adjacent' *Base* $\cos = \frac{B}{H}$
To Plaster Back side AB is called 'hypotenuse'. *Hypot* $\tan = \frac{P}{B}$

Then

$$\sin \theta = \frac{\text{opposite}}{\text{hypotenuse}}, \quad \cos \theta = \frac{\text{adjacent}}{\text{hypotenuse}}, \quad \tan \theta = \frac{\text{opposite}}{\text{adjacent}}$$

$$\operatorname{cosec} \theta = \frac{\text{hypotenuse}}{\text{opposite}}, \quad \sec \theta = \frac{\text{hypotenuse}}{\text{adjacent}}, \quad \cot \theta = \frac{\text{adjacent}}{\text{opposite}}$$

Thus it is seen that the last three functions are merely the reciprocals of the first three, i.e.

$$\operatorname{cosec} \theta = \frac{1}{\sin \theta} \quad \sec \theta = \frac{1}{\cos \theta} \quad \cot \theta = \frac{1}{\tan \theta}.$$

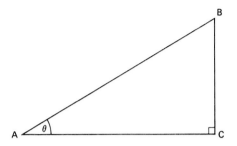

Fig. 12.1   A right-angled triangle with acute angle θ (see page 218).

Also, it is easy to verify that

$$\tan \theta = \frac{\sin \theta}{\cos \theta} \qquad \cot \theta = \frac{\cos \theta}{\sin \theta}.$$

Defining these six trigonometric functions in relation to a right-angled triangle restricts consideration of them to values of θ in the range $0° \leqslant \theta \leqslant 90°$. There is no reason, however, to prevent us from considering a trigonometric function of θ, when θ exceeds $90°$, or even where θ is a 'negative' angle; and the second section of this chapter will be devoted to such general angles and their trigonometric functions.

For the moment, however, let us assume that we can plot a graph of $\sin \theta$ against θ for the range $0° \leqslant \theta \leqslant 720°$; the curve is shown in Fig. 12.2. We see that the portion of the curve from $0°$ to $360°$ is repeated exactly

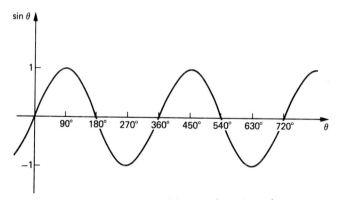

Fig. 12.2   Curve of the function $f(\theta) = \sin \theta$ (θ in degrees).

from 360° to 720°, and indeed, the curve will continue to be repeated every 360°. The curve closely resembles a waveform; the wavelength is 360°, and the amplitude is 2 (from −1 to +1). This curve is so well known that any curve of this type is referred to as a 'sine curve', and probably its most important property is the exact repetition every 360°.

All trigonometric functions are cyclic in this way, and this property, together with the fact that they are associated with angular measure, makes for their practical importance. In biology, trigonometric functions arise when considering the radiation environment of plants and animals; days and nights are a cyclic phenomenon, and the intensity of radiation on a surface involves the use of angular measure. Population dynamics, which can only be satisfactorily studied by the use of mathematics, often needs to explain why some populations oscillate, and this naturally leads to trigonometric functions: an example of this is given in Chapter 13, page 252. Again, the study of muscle movement in animals involves trigonometric functions. It is therefore very useful for the quantitative biologist to study trigonometric functions; but their theory and use is not as straightforward as for those functions considered hitherto, and so this chapter is best deferred to a second reading of this book.

THE GENERAL ANGLE

Radian measure

In Fig. 12.2 is shown the graph of the function $\sin \theta$ plotted against θ. We could, if we wished, put $y = \sin \theta$ and then the vertical axis would be y, and the graph would be that of the function $y = \sin \theta$. Now when drawing graphs of other functions, e.g. $y = ae^{bx}$, both axes have been a scale of natural numbers. In the present case, however, although the vertical axis is a scale of natural numbers, the horizontal axis is not. The scale of degrees is quite arbitrary, but has historical significance. The basis of degrees as an angular measure is that a complete circle subtends an angle of 360° at its centre, and so a right-angle, which is the angle subtended by a quarter of a circle, is 360°/4 = 90°. The question then immediately arises, 'is it possible to have a more natural angular measure?'

What needs to be done is to have another scale of angle measurement whose unit is defined by some definite (i.e. non-arbitrary) quantity. Consider a circle of radius r (Fig. 12.3) and centre O; A and B are two points on the circle such that the length of the arc between them is equal to the radius of the circle, r. Then angle $A\hat{O}B$ is defined as unity on the natural scale of angular measure, and the unit is known as a radian (rad).

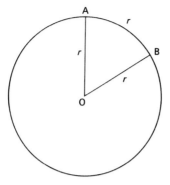

Fig. 12.3   A sector of a circle in which the arc is equal to the radius. The angle AÔB is 1 radian.

Thus

$$A\hat{O}B = 1 \text{ radian.}$$

Now we need to know the relationship between degrees and radians. The arc **AB** is part of the circumference of the circle, is equal to the radius, r, and subtends an angle of 1 rad at the centre. The whole circumference is $2\pi r$ and this subtends an angle of $360°$ at the centre. Hence if

$$r \text{ (of circumference) subtends 1 rad}$$

then

$$2\pi r \text{ (of circumference) subtends } (2\pi) \text{ rad}$$

So

$$(2\pi) \text{ rad} = 360°$$

hence

$$1 \text{ rad} = \frac{360°}{2\pi} = \frac{180°}{\pi} \simeq 57°18' \qquad (12.1)$$

and

$$1° = \left(\frac{2\pi}{360}\right) \text{ rad} = \left(\frac{\pi}{180}\right) \text{ rad} \simeq 0.01745 \text{ rad} \qquad (12.2)$$

Since $360° = (2\pi)$ rad, we have $180° = \pi$ rad, $90° = (\pi/2)$ rad, $45° = (\pi/4)$ rad, $270° = (3\pi/2)$ rad, and so on. In Fig. 12.2, if the horizontal axis is measured in radians instead of degrees, then the sine curve will appear as in Fig. 12.4.

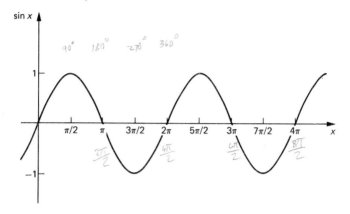

Fig. 12.4   Curve of the function $f(x) = \sin x$ (x in radians).

Now, both axes may be regarded as natural numbers; notice that we have re-labelled the horizontal axis x rather than θ. Although in many cases it does not matter whether angles are measured in degrees or radians, it is preferable to use radians when in doubt; and it is conventional to designate a scale of angles in a small Roman letter when measured in radians, and a Greek letter when the angle is measured in degrees.

The general angle and its trigonometric functions

Consider a circle whose centre is placed at the origin of x–y axes, and whose radius is unity (Fig. 12.5). Let there also be a point P on the circumference, and let the straight line OP subtend an angle θ with the x-axis, measured in an *anti-clockwise* direction.

$0° < \theta < 90°$ *(0 rad $< x <$ π/2 rad)*

Now if θ is acute, point P is in the upper right-hand quadrant of the co-ordinate axes, as shown in Fig. 12.5(a). The triangle OPQ is right-angled at Q and so from previous definitions (page 218)

$$\sin \theta = \frac{PQ}{OP} = \frac{PQ}{1} = PQ \quad \text{(a measurement on the } y\text{-axis)}$$

$$\cos \theta = \frac{OQ}{OP} = \frac{OQ}{1} = OQ \quad \text{(a measurement on the } x\text{-axis)}$$

$$\tan \theta = \frac{PQ}{OQ} \qquad \text{(ratio of measurement on the } y\text{-axis to the measurement on the } x\text{-axis)}$$

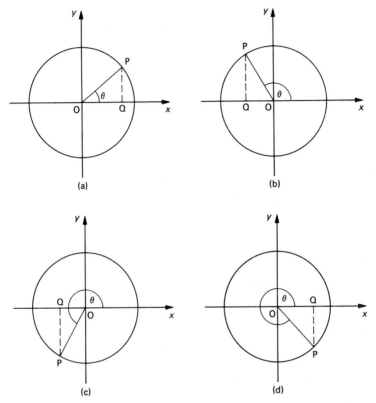

Fig. 12.5 Circle of unit radius, centred on the origin of the co-ordinate axes, defining the trigonometrical function of angle θ: (a) $0° < \theta < 90°$, (b) $90° < \theta < 180°$, (c) $180° < \theta < 270°$, (d) $270° < \theta < 360°$.

Thus, the definitions of the trigonometrical functions can be re-stated in terms of a circle whose centre is at the origin of the usual co-ordinate axes, and in this way we may define these functions for angles in excess of $90°$, $(\pi/2)$ rad. In the following, we shall only deal with the sine, cosine, and tangent. The remaining three functions may be obtained from these quite easily, as reciprocals (page 218).

Before leaving the upper right-hand quadrant, we note that

$$\sin 0° = 0 \quad (PQ = 0) \qquad \sin 90° = 1 \quad (PQ = 1)$$
$$\cos 0° = 1 \quad (OQ = 1) \qquad \cos 90° = 0 \quad (OQ = 0)$$
$$\tan 0° = 0 \quad (PQ/OQ = 0) \qquad \tan 90° \text{ is infinitely large.}$$

These results can be checked in the usual tables of trigonometric functions, and these tables also give the values of sine, cosine, and tangent of any angle lying in the range $0° < \theta < 90°$.

$90° < \theta < 180°$ ($\pi/2$ rad $< x < \pi$ rad)
sin θ: Since PQ is positive (Fig. 12.5(b)), sin θ is positive in this range. The length of OP is *always* positive. Now sin $90° = 1$ (PQ = 1), and sin $180° = 0$ (PQ = 0).
So

$$\sin \theta = \sin(180 - \theta)$$

cos θ: Since OQ is negative, cos θ is negative in this range. Now cos $90° = 0$ (OQ = 0), and cos $180° = -1$ (OQ = -1).
So

$$\cos \theta = -\cos(180 - \theta)$$

tan θ:

$$\tan \theta = \frac{\sin \theta}{\cos \theta} = \frac{\sin(180 - \theta)}{-\cos(180 - \theta)}$$

So

$$\tan \theta = -\tan(180 - \theta)$$

$180° < \theta < 270°$ (π rad $< x < 3\pi/2$ rad)
sin θ: Since PQ is negative (Fig. 12.5(c)), sin θ is negative in this range. Now sin $180° = 0$ (PQ = 0), and sin $270° = -1$ (PQ = -1).
So

$$\sin \theta = -\sin(\theta - 180)$$

cos θ: Since OQ is negative, cos θ is negative in this range. Now cos $180° = -1$ (OQ = -1), and cos $270° = 0$ (OQ = 0).
So

$$\cos \theta = -\cos(\theta - 180)$$

tan θ:

$$\tan \theta = \frac{\sin \theta}{\cos \theta} = \frac{-\sin(\theta - 180)}{-\cos(\theta - 180)}$$

So

$$\tan \theta = \tan(\theta - 180)$$

$270° < \theta < 360°$ ($3\pi/2$ rad $< x < 2\pi$ rad)

sin θ: Since PQ is negative (Fig. 12.6(d)), sin θ is negative in this range. Now sin $270° = -1$ (PQ $= -1$), and sin $360° = 0$ (PQ $= 0$). So

$$\sin \theta = -\sin(360 - \theta)$$

cos θ: Since OQ is positive, cos θ is positive in this range. Now cos $270° = 0$ (OQ $= 0$), and cos $360° = 1$ (OQ $= 1$). So

$$\cos \theta = \cos(360 - \theta)$$

tan θ:

$$\tan \theta = \frac{\sin \theta}{\cos \theta} = \frac{-\sin(360 - \theta)}{\cos(360 - \theta)}$$

So

$$\tan \theta = -\tan(360 - \theta)$$

$\theta > 360°$ ($x > 2\pi$ rad)

If $\theta > 360°$ this means that point P has completed at least one revolution around the circle in an anti-clockwise direction. If $360° < \theta \leqslant 720°$, the trigonometric functions of θ can be found by first deducting $360°$ from θ, and then by using the above rules. So

$$\sin \theta = \sin(\theta - 360)$$
$$\cos \theta = \cos(\theta - 360)$$
$$\tan \theta = \tan(\theta - 360)$$

If $720° < \theta \leqslant 1080°$ then e.g. sin $\theta = \sin(\theta - 720)$, and so on. It is thus evident that trigonometric functions repeat themselves every $360°$, or, more important, every 2π radians.

Negative angles

These are obtained by rotating P in a clockwise direction. So if $0° > \theta \geqslant -360°$, we have, for example

$$\sin \theta = \sin(\theta + 360)$$

If $-360° > \theta \geqslant -720°$, we have, for example

$$\tan \theta = \tan(\theta + 720)$$

and so on.

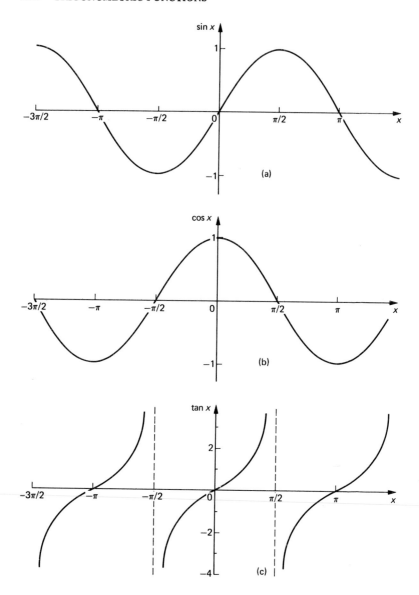

Fig. 12.6   The trigonometrical functions of x over the range $-3\pi/2 \leqslant x \leqslant 3\pi/2$: (a) sin x, (b) cos x, (c) tan x, (d) cosec x, (e) sec x, (f) cot x.

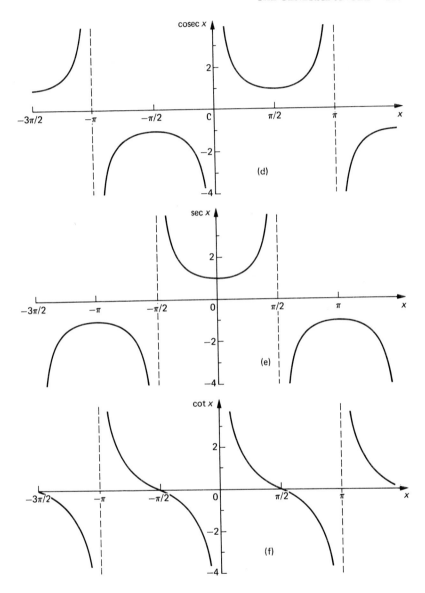

(d)

(e)

(f)

Thus, although trigonometric functions are only tabulated for the range $0° \leqslant \theta \leqslant 90°$, they can be found for any angle by making use of the above rules.

It is also apparent that the trigonometric functions are defined in relation to a circle, and that their definition in connexion with a triangle is a special case. Because of their association with a circle, trigonometric functions are sometimes called **circular functions**.

EXAMPLE 12.1
Evaluate

(a) sin 215°   (b) cos 321°   (c) tan 160°   (d) sin 399°   (e) cos 540°
(f) tan 611°   (g) cos −72°   (h) tan −300°

(a) $180° < \theta < 270°$:

hence

$$\sin 215° = -\sin(215 - 180) = -\sin 35° = -0.5736$$

(b) $270° < \theta < 360°$:

hence

$$\cos 321° = \cos(360 - 321) = \cos 39° = 0.7771$$

(c) $90° < \theta < 180°$:

hence

$$\tan 160° = -\tan(180 - 160) = -\tan 20° = -0.3640$$

(d) $360° < \theta < 720°$:

hence

$$\sin 399° = \sin(399 - 360) = \sin 39° = 0.6293$$

(e) $360° < \theta < 720°$:

hence

$$\cos 540° = \cos(540 - 360) = \cos 180° = -1.0000$$

(f) $360° < \theta < 720°$:

hence

$$\tan 611° = \tan(611 - 360) = \tan 251°$$
$$180° < \theta < 270°:$$

hence

$$\tan 251° = \tan(251 - 180) = \tan 71° = 2.904$$

thus

$$\tan 611° = 2.904$$

(g) $0° < \theta < -360°:$

hence

$$\cos -72° = \cos(-72 + 360) = \cos 288°$$
$$270° < \theta < 360°:$$

hence

$$\cos 288° = \cos(360 - 288) = \cos 72° = 0.3090$$

thus

$$\cos -72° = 0.3090$$

(h) $0° < \theta < -360°:$

hence

$$\tan -300° = \tan(-300 + 360) = \tan 60° = 1.732$$

thus

$$\tan -300° = 1.732$$

Graphs of all six trigonometric functions are shown in Fig. 12.6.

TRIGONOMETRIC IDENTITIES

Before dealing with the identities in this section, it is necessary to consider an item of notation. If (say) $\sin x$ is raised to the power n, that is $(\sin x)^n$, it is customary to write it as $\sin^n x$. Thus

$$\sin^2 x = (\sin x)^2 = (\sin x)(\sin x)$$

Distinguish carefully between $\sin^2 x$ and $\sin x^2$, the latter is

$$\sin x^2 = \sin(x)^2 = \sin(x \cdot x)$$

Identities involving a single angle

In relation to the circle in Fig. 12.5, we have shown that $\sin \theta = PQ$ and $\cos \theta = OQ$ (page 222). Now the right-angled triangle OPQ, by Pythagoras'

Theorem

$$OQ^2 + PQ^2 = OP^2$$

But $OQ = \cos\theta$, $PQ = \sin\theta$, and $OP = 1$ (radius of the circle). Hence

$$\cos^2\theta + \sin^2\theta = 1 \qquad (12.3)$$

This is the **fundamental identity** in trigonometry; and two other useful ones can be derived from it. Divide both sides of 12.3 by $\cos^2\theta$:

$$1 + \frac{\sin^2\theta}{\cos^2\theta} = \frac{1}{\cos^2\theta}$$

i.e.

$$1 + \tan^2\theta = \sec^2\theta \qquad (12.4)$$

Again, divide both sides of 12.3 by $\sin^2\theta$:

$$\frac{\cos^2\theta}{\sin^2\theta} + 1 = \frac{1}{\sin^2\theta}$$

i.e.

$$\cot^2\theta + 1 = \operatorname{cosec}^2\theta \qquad (12.5)$$

Because they have been derived from the circular definition of the trigonometric functions, these identities are valid for all values of θ.

Identities of compound angles

Consider two angles, θ and ϕ; then we may define trigonometric functions of the two angles added together or subtracted from one another, e.g. $\sin(\theta + \phi)$, $\tan(\theta - \phi)$. However, it is frequently necessary to be able to express such quantities in terms of a combination of trigonometric functions of each individual angle. For instance, it will be shown below that $\sin(\theta + \phi) = \sin\theta \cdot \cos\phi + \cos\theta \cdot \sin\phi$. This kind of expansion of the trigonometric functions of compound angles is very important in mathematics generally, but for the biologist, interest in such formulae lies in their use in deriving the differential coefficients of trigonometric functions. We shall prove the results for acute angles, but they are valid for all angles.

In Fig. 12.7, let $M\widehat{O}N = \theta$ and $M\widehat{O}P = \phi$. Draw PM such that $P\widehat{M}O = 90°$, and draw PL such that $P\widehat{L}O = 90°$; also draw MS such that $M\widehat{S}P = 90°$, so that LNMS is a rectangle. Since PL and PM are perpendicular to ON and OM, respectively, $M\widehat{P}S = \theta$.

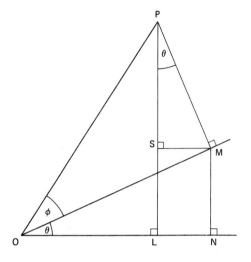

Fig. 12.7   Diagram to derive the trigonometric functions of the compound angle $(\theta \pm \phi)$.

Now

$$\sin(\theta + \phi) = \frac{LP}{OP} = \frac{LS + PS}{OP} = \frac{MN}{OP} + \frac{PS}{OP}$$

$$= \frac{MN}{MO} \cdot \frac{MO}{OP} + \frac{PS}{MP} \cdot \frac{MP}{OP}$$

i.e.

$$\sin(\theta + \phi) = \sin \theta \cdot \cos \phi + \cos \theta \cdot \sin \phi \qquad (12.6)$$

$$\cos(\theta + \phi) = \frac{LO}{OP} = \frac{NO - LN}{OP} = \frac{NO}{OP} - \frac{MS}{OP}$$

$$= \frac{NO}{MO} \cdot \frac{MO}{OP} - \frac{MS}{MP} \cdot \frac{MP}{OP}$$

i.e.

$$\cos(\theta + \phi) = \cos \theta \cdot \cos \phi - \sin \theta \cdot \sin \phi \qquad (12.7)$$

$$\tan(\theta + \phi) = \frac{\sin(\theta + \phi)}{\cos(\theta + \phi)} = \frac{\sin \theta \cdot \cos \phi + \cos \theta \cdot \sin \phi}{\cos \theta \cdot \cos \phi - \sin \theta \cdot \sin \phi}$$

Divide numerator and denominator of the right-hand side by $\cos \theta \cdot \cos \phi$; then we have

$$\tan(\theta + \phi) = \frac{\dfrac{\sin \theta \cdot \cos \phi}{\cos \theta \cdot \cos \phi} + \dfrac{\cos \theta \cdot \sin \phi}{\cos \theta \cdot \cos \phi}}{\dfrac{\cos \theta \cdot \cos \phi}{\cos \theta \cdot \cos \phi} - \dfrac{\sin \theta \cdot \sin \phi}{\cos \theta \cdot \cos \phi}}$$

$$\tan(\theta + \phi) = \frac{\tan \theta + \tan \phi}{1 - \tan \theta \cdot \tan \phi} \qquad (12.8)$$

Now put $\phi = -\phi$ in 12.6, then

$$\sin\{\theta + (-\phi)\} = \sin \theta \cdot \cos(-\phi) + \cos \theta \cdot \sin(-\phi)$$

But $\theta + (-\phi) = \theta - \phi$, and from the definitions of the general angle given earlier (page 223), $\cos(-\phi) = \cos \phi$, $\sin(-\phi) = -\sin \phi$. So

$$\sin(\theta - \phi) = \sin \theta \cdot \cos \phi - \cos \theta \cdot \sin \phi \qquad (12.9)$$

Put $\phi = -\phi$ in 12.7, then

$$\cos\{\theta + (-\phi)\} = \cos \theta \cdot \cos(-\phi) - \sin \theta \cdot \sin(-\phi)$$

so

$$\cos(\theta - \phi) = \cos \theta \cdot \cos \phi + \sin \theta \cdot \sin \phi \qquad (12.10)$$

Put $\phi = -\phi$ in 12.8. Now $\tan(-\phi) = -\tan \phi$

So

$$\tan(\theta - \phi) = \frac{\tan \theta - \tan \phi}{1 + \tan \theta \cdot \tan \phi} \qquad (12.11)$$

If $\phi = \theta$ in 12.6, 12.7, and 12.8, we have

$$\sin(\theta + \theta) = \sin \theta \cdot \cos \theta + \cos \theta \cdot \sin \theta$$

i.e.

$$\sin 2\theta = 2 \cdot \sin \theta \cdot \cos \theta \qquad (12.12)$$

$$\cos(\theta + \theta) = \cos \theta \cdot \cos \theta - \sin \theta \cdot \sin \theta$$

i.e.

$$\cos 2\theta = \cos^2 \theta - \sin^2 \theta \qquad (12.13)$$

$$\tan(\theta + \theta) = \frac{\tan \theta + \tan \theta}{1 - \tan \theta \cdot \tan \theta}$$

i.e.

$$\tan 2\theta = \frac{2 \cdot \tan \theta}{1 - \tan^2 \theta} \qquad (12.14)$$

Finally, we develop four other results which will be useful later. Adding together equations 12.6 and 12.9 gives

$$\sin(\theta + \phi) = \sin \theta \cdot \cos \phi + \cos \theta \cdot \sin \phi$$
$$\text{plus } \sin(\theta - \phi) = \sin \theta \cdot \cos \phi - \cos \theta \cdot \sin \phi$$

so

$$\sin(\theta + \phi) + \sin(\theta - \phi) = 2 \cdot \sin \theta \cdot \cos \phi \qquad (12.15)$$

Equation 12.6 minus equation 12.9 gives

$$\sin(\theta + \phi) = \sin \theta \cdot \cos \phi + \cos \theta \cdot \sin \phi$$
$$\text{minus } \sin(\theta - \phi) = \sin \theta \cdot \cos \phi - \cos \theta \cdot \sin \phi$$

so

$$\sin(\theta + \phi) - \sin(\theta - \phi) = 2 \cdot \cos \theta \cdot \sin \phi \qquad (12.16)$$

Adding equations 12.7 and 12.10 gives

$$\cos(\theta + \phi) = \cos \theta \cdot \cos \phi - \sin \theta \cdot \sin \phi$$
$$\text{plus } \cos(\theta - \phi) = \cos \theta \cdot \cos \phi + \sin \theta \cdot \sin \phi$$

so

$$\cos(\theta + \phi) + \cos(\theta - \phi) = 2 \cdot \cos \theta \cdot \cos \phi \qquad (12.17)$$

Equation 12.7 minus equation 12.10 gives

$$\cos(\theta + \phi) = \cos \theta \cdot \cos \phi - \sin \theta \cdot \sin \phi$$
$$\text{minus } \cos(\theta - \phi) = \cos \theta \cdot \cos \phi + \sin \theta \cdot \sin \phi$$

so

$$\cos(\theta + \phi) - \cos(\theta - \phi) = -2 \cdot \sin \theta \cdot \sin \phi \qquad (12.18)$$

THE CALCULUS OF TRIGONOMETRIC FUNCTIONS

Limit theorems

In Chapter 5, which introduced the differential calculus, we discussed limits before differentiation. Before the trigonometric functions can be differentiated, special trigonometric limits need to be presented. Only the first of these is essential for our purpose, but two allied ones are given for the sake of completeness.

All angles involved in limits, and in the calculus, of trigonometric functions are measured in radians.

Theorem 12.1.
$$\lim_{x \to 0} \left(\frac{\sin x}{x} \right) = 1$$

where the angle x is measured in radians.

In Fig. 12.8, AB is an arc of a circle, centre O and radius r, subtending an angle x ($<\pi/2$ rad) at O. The line BT is a tangent to the circle at B, extended to meet OA, produced, at T. The line BN is drawn such that BNO is a right-angle. Now inspection of Fig. 12.8 shows that

$$BN < AB < BT$$

Divide throughout by BO:

$$\frac{BN}{BO} < \frac{AB}{BO} < \frac{BT}{BO}$$

i.e.

$$\sin x < x < \tan x \tag{12.19}$$

(since x is in radians).

Invert throughout, and replace $\tan x$ by $(\sin x)/(\cos x)$

$$\frac{1}{\sin x} > \frac{1}{x} > \frac{\cos x}{\sin x}$$

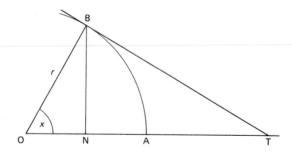

Fig. 12.8   Diagram to investigate the limit of $(\sin x)/x$ as x approaches 0.

and multiplying through by sin x gives

$$1 > \frac{\sin x}{x} > \cos x$$

Thus $(\sin x)/x$ lies between 1 and $\cos x$. But as $x \to 0$, $\cos x \to 1$; hence $(\sin x)/x$ lies between 1 and a quantity less than 1, but which approaches 1 as $x \to 0$. Therefore

$$\lim_{x \to 0} \left(\frac{\sin x}{x} \right) = 1 \qquad (12.20)$$

Theorem 12.2. $\qquad \lim_{x \to 0} \left(\frac{\tan x}{x} \right) = 1$

where the angle x is measured in radians.

Divide through the inequality 12.19 by $\tan x$:

$$\cos x < \frac{x}{\tan x} < 1$$

Invert throughout,

$$\sec x > \frac{\tan x}{x} > 1$$

Thus $(\tan x)/x$ lies between $\sec x$ and 1. But as $x \to 0$, $\sec x \to 1$; hence $(\tan x)/x$ lies between 1 and a quantity greater than 1, but which approaches 1 as $x \to 0$. Therefore

$$\lim_{x \to 0} \left(\frac{\tan x}{x} \right) = 1 \qquad (12.21)$$

These two results show that, if x is very small and is measured in radians, then sin x can be replaced by x, and also tan x can be replaced by x. The smaller the value of x, the more accurate will be the results. Tables show that there is a fair measure of agreement between x, sin x, and tan x, for angles up to about 0.08727 radians (5°).

Theorem 12.3. $\qquad \lim_{x \to 0} \left(\frac{1 - \cos x}{x^2} \right) = \frac{1}{2}$

where the angle x is measured in radians.

Replace θ by $\frac{1}{2}x$ in identity 12.13, then we have

$$\cos x = \cos^2 \tfrac{1}{2}x - \sin^2 \tfrac{1}{2}x \qquad (12.22)$$

From identity 12.3, we have

$$\cos^2 \tfrac{1}{2}x = 1 - \sin^2 \tfrac{1}{2}x$$

So, substituting for $\cos^2 \tfrac{1}{2}x$ in 12.22 yields

$$\cos x = 1 - \sin^2 \tfrac{1}{2}x - \sin^2 \tfrac{1}{2}x$$

i.e.

$$1 - \cos x = 2 \cdot \sin^2 \tfrac{1}{2}x$$

Hence

$$\frac{1 - \cos x}{x^2} = \frac{2 \cdot \sin^2 \tfrac{1}{2}x}{x^2} = 2\left(\frac{\sin \tfrac{1}{2}x}{x}\right)^2$$

and so

$$\frac{1 - \cos x}{x^2} = \frac{1}{2}\left(\frac{\sin \tfrac{1}{2}x}{\tfrac{1}{2}x}\right)^2$$

Since

$$\lim_{x \to 0}\left(\frac{\sin \tfrac{1}{2}x}{\tfrac{1}{2}x}\right) = 1$$

then

$$\lim_{x \to 0}\left(\frac{1 - \cos x}{x^2}\right) = \tfrac{1}{2} \times (1)^2 = \tfrac{1}{2}. \tag{12.23}$$

Differention

We shall differentiate $\sin x$ and $\cos x$ from first principles, and then use these results combined with the various rules given in Chapter 5 to differentiate the remaining trigonometric functions.

EXAMPLE 12.2

Find

(a) $\dfrac{d(\sin x)}{dx}$ (b) $\dfrac{d(\cos x)}{dx}$ (c) $\dfrac{d(\tan x)}{dx}$ (d) $\dfrac{d(\csc x)}{dx}$

(e) $\dfrac{d(\sec x)}{dx}$ (f) $\dfrac{d(\cot x)}{dx}$

(a) Put $y = \sin x$ (12.24)

Let there be a small increase of δx in x and a corresponding increment of

δy in y; then

$$y + \delta y = \sin(x + \delta x) \qquad (12.25)$$

Subtracting equation 12.24 from equation 12.25 gives

$$\delta y = \sin(x + \delta x) - \sin x \qquad (12.26)$$

Now we use identity 12.16, with $\theta = x + \frac{1}{2}\delta x$ and $\phi = \frac{1}{2}\delta x$.
This gives

$$\sin\{(x + \tfrac{1}{2}\delta x) + \tfrac{1}{2}\delta x\} - \sin\{(x + \tfrac{1}{2}\delta x) - \tfrac{1}{2}\delta x\} = 2 \cdot \cos(x + \tfrac{1}{2}\delta x) \cdot \sin \tfrac{1}{2}\delta x$$

i.e.

$$\sin(x + \delta x) - \sin x = 2 \cdot \cos(x + \tfrac{1}{2}\delta x) \cdot \sin \tfrac{1}{2}\delta x$$

Substituting into the right-hand side of 12.26, we have

$$\delta y = 2 \cdot \cos(x + \tfrac{1}{2}\delta x) \cdot \sin \tfrac{1}{2}\delta x$$

Divide both sides by δx:

$$\frac{\delta y}{\delta x} = \frac{2 \cdot \cos(x + \tfrac{1}{2}\delta x) \cdot \sin \tfrac{1}{2}\delta x}{\delta x}$$

Divide both numerator and denominator of the right-hand side by 2:

$$\frac{\delta y}{\delta x} = \cos(x + \tfrac{1}{2}\delta x) \cdot \frac{\sin \tfrac{1}{2}\delta x}{\tfrac{1}{2}\delta x}$$

So

$$\frac{dy}{dx} = \lim_{\delta x \to 0}\left\{ \cos(x + \tfrac{1}{2}\delta x) \cdot \frac{\sin \tfrac{1}{2}\delta x}{\tfrac{1}{2}\delta x} \right\} = \cos x$$

since

$$\lim_{\delta x \to 0}\left(\frac{\sin \tfrac{1}{2}\delta x}{\tfrac{1}{2}\delta x} \right) = 1 \quad \text{(page 234).}$$

Thus

$$\frac{d(\sin x)}{dx} = \cos x \qquad (12.27)$$

(b) Put $$y = \cos x \qquad (12.28)$$

Let there be a small increase of δx in x and a corresponding increment of δy in y; then

$$y + \delta y = \cos(x + \delta x) \qquad (12.29)$$

Subtracting equation 12.28 from 12.29

$$\delta y = \cos(x + \delta x) - \cos x$$

Beginning with the identity 12.18, we obtain in a similar manner to the previous example

$$\delta y = -2 \cdot \sin(x + \tfrac{1}{2}\delta x) \cdot \sin \tfrac{1}{2}\delta x$$

Dividing both sides by δx, and then the right-hand side by 2,

$$\frac{\delta y}{\delta x} = -\sin(x + \tfrac{1}{2}\delta x) \cdot \frac{\sin \tfrac{1}{2}\delta x}{\tfrac{1}{2}\delta x}$$

$$\frac{dy}{dx} = \lim_{\delta x \to 0} \left\{ -\sin(x + \tfrac{1}{2}\delta x) \cdot \frac{\sin \tfrac{1}{2}\delta x}{\tfrac{1}{2}\delta x} \right\} = -\sin x$$

Thus

$$\frac{d(\cos x)}{dx} = -\sin x \tag{12.30}$$

(c)
$$\tan x = \frac{\sin x}{\cos x}$$

Hence, applying the quotient rule, we have

$$\frac{d(\tan x)}{dx} = \frac{(\cos x) \cdot \dfrac{d(\sin x)}{dx} - (\sin x) \cdot \dfrac{d(\cos x)}{dx}}{\cos^2 x}$$

Therefore

$$\frac{d(\tan x)}{dx} = \frac{\cos^2 x + \sin^2 x}{\cos^2 x} = \frac{1}{\cos^2 x}$$

(using the fundamental identity, 12.3), and so

$$\frac{d(\tan x)}{dx} = \sec^2 x \tag{12.31}$$

(d)
$$\operatorname{cosec} x = \frac{1}{\sin x}$$

Applying the quotient rule, we have

$$\frac{d(\operatorname{cosec} x)}{dx} = \frac{-\cos x}{\sin^2 x} = -\frac{1}{\sin x} \cdot \frac{\cos x}{\sin x}$$

Thus

$$\frac{d(\operatorname{cosec} x)}{dx} = -\operatorname{cosec} x \cdot \cot x \qquad (12.32)$$

(e)
$$\sec x = \frac{1}{\cos x}$$

Again, one could apply the quotient rule, but for a change this example will be worked by the function of a function rule. Put $y = \sec x$, i.e. $y = 1/\cos x$ and put $u = \cos x$, so that

$$y = \frac{1}{u}.$$

Then

$$\frac{du}{dx} = -\sin x \quad \text{and} \quad \frac{dy}{du} = -\frac{1}{u^2}$$

Thus

$$\frac{dy}{dx} = \left(-\frac{1}{u^2}\right)\left(-\sin x\right) = \frac{\sin x}{\cos^2 x} = \frac{1}{\cos x} \cdot \frac{\sin x}{\cos x}$$

and so

$$\frac{d(\sec x)}{dx} = \sec x \cdot \tan x \qquad (12.33)$$

(f)
$$\cot x = \frac{\cos x}{\sin x}$$

Applying the quotient rule:

$$\frac{d(\cot x)}{dx} = \frac{(\sin x) \cdot \dfrac{d(\cos x)}{dx} - (\cos x) \cdot \dfrac{d(\sin x)}{dx}}{\sin^2 x}$$

$$= \frac{-\sin^2 x - \cos^2 x}{\sin^2 x} = -\frac{(\sin^2 x + \cos^2 x)}{\sin^2 x}$$

So, using the fundamental identity, 12.3, we finally have

$$\frac{d(\cot x)}{dx} = -\operatorname{cosec}^2 x \qquad (12.34)$$

More complex trigonometric functions can be differentiated by the usual rules applied to the above basic results.

EXAMPLE 12.3

Differentiate with respect to x:

(a) $\sin^3 x$ (b) $\cos x^4$ (c) $\sec(a + bx)$

(a) Put $y = \sin^3 x$ and $u = \sin x,$ then $y = u^3$

so

$$\frac{du}{dx} = \cos x$$

and

$$\frac{dy}{du} = 3u^2$$

Hence

$$\frac{dy}{dx} = 3u^2 \cdot \cos x$$

and so

$$\frac{d(\sin^3 x)}{dx} = 3 \cdot \sin^2 x \cdot \cos x$$

(b) Put $y = \cos x^4$ and $u = x^4,$ then $y = \cos u$

so

$$\frac{du}{dx} = 4x^3 \quad \text{and} \quad \frac{dy}{du} = -\sin u$$

Hence

$$\frac{dy}{dx} = (-\sin u) \cdot 4x^3$$

and so

$$\frac{d(\cos x^4)}{dx} = -4x^3 \cdot \sin x^4$$

(c) Put $y = \sec(a + bx)$ and $u = a + bx,$ then $y = \sec u$

so

$$\frac{du}{dx} = b$$

and

$$\frac{dy}{du} = \sec u \cdot \tan u$$

$$\frac{dy}{dx} = b \cdot \sec u \cdot \tan u$$

i.e.

$$\frac{d\{\sec(a + bx)\}}{dx} = b \cdot \sec(a + bx) \cdot \tan(a + bx)$$

Integration

Reversal of the equations 12.27, and 12.30 to 12.34 inclusive, implies that the integrals of the right-hand sides of these equations are equal to the left-hand sides. In particular, we have:

$$\int (\cos x)dx = \sin x + c \qquad (12.35)$$

and

$$\int (\sin x)dx = -\cos x + c \qquad (12.36)$$

These two results are fundamental to the integration of trigonometric functions, and integrals of the remaining functions are found by applying various methods to the two standard integrals 12.35 and 12.36.

EXAMPLE 12.4
Find

(a) $\int (\tan x)dx$ (b) $\int (\operatorname{cosec} x)dx$ (c) $\int (\sec x)dx$ (d) $\int (\cot x)dx$

(a) $$\int (\tan x)dx = \int \frac{\sin x}{\cos x} dx$$

Make the substitution $u = \cos x$, then

$$\frac{du}{dx} = -\sin x$$

and so

$$\sin x = -\frac{du}{dx} \cdot$$

Therefore

$$\int \frac{(\sin x)dx}{\cos x} = \int \frac{-\dfrac{du}{dx}dx}{u} = -\int \frac{du}{u} = -\log_e u + c = -\log_e(\cos x) + c$$

Now

$$-\log_e(\cos x) = \log_e(1/\cos x)$$

so

$$\int (\tan x)dx = \log_e(\sec x) + c \qquad (12.37)$$

(b)
$$\int (\operatorname{cosec} x)dx = \int \frac{dx}{\sin x}$$

Now from 12.12

$$\sin x = 2 \cdot \sin \frac{1}{2}x \cdot \cos \frac{1}{2}x$$

so

$$\int \frac{dx}{\sin x} = \frac{1}{2} \int \frac{dx}{\sin \frac{1}{2}x \cdot \cos \frac{1}{2}x}$$

Divide numerator and denominator of the right-hand side by $\cos^2 \frac{1}{2}x$; then

$$\frac{1}{2} \int \frac{\dfrac{dx}{\cos^2 \frac{1}{2}x}}{\dfrac{\sin \frac{1}{2}x}{\cos \frac{1}{2}x} \cdot \dfrac{\cos \frac{1}{2}x}{\cos \frac{1}{2}x}} = \frac{1}{2} \int \frac{\sec^2 \frac{1}{2}x}{\tan \frac{1}{2}x} dx$$

Put $u = \tan \frac{1}{2}x$, then

$$\frac{du}{dx} = \frac{1}{2} \cdot \sec^2 \frac{1}{2}x;$$

i.e.

$$dx = \frac{2du}{\sec^2 \frac{1}{2}x}$$

Hence

$$\frac{1}{2} \int \frac{(\sec^2 \frac{1}{2}x)dx}{\tan \frac{1}{2}x} = \frac{1}{2} \int \frac{\sec^2 \frac{1}{2}x \cdot \dfrac{2du}{\sec^2 \frac{1}{2}x}}{u} = \int \frac{du}{u} = \log_e u + c$$

and so

$$\int(\operatorname{cosec} x)dx = \log_e(\tan \tfrac{1}{2}x) + c \qquad (12.38)$$

(c) Multiply both numerator and denominator by $(\sec x + \tan x)$, then

$$\int(\sec x)dx = \int \frac{\sec x \cdot (\sec x + \tan x)}{\sec x + \tan x} dx \qquad (12.39)$$

Now make substitution $u = \sec x + \tan x$, then

$$\frac{du}{dx} = \sec x \cdot \tan x + \sec^2 x$$

$$= \sec x \cdot (\sec x + \tan x)$$

$$= u \cdot \sec x$$

Hence

$$dx = \frac{du}{u \cdot \sec x}$$

Substituting into the right-hand side of 12.39 gives

$$\int(\sec x)dx = \int \frac{(\sec x) \cdot u \cdot \dfrac{du}{u \cdot \sec x}}{u} = \int \frac{du}{u} = \log_e u + c$$

and so

$$\int(\sec x)dx = \log_e(\sec x + \tan x) + c \qquad (12.40)$$

(d) $$\int(\cot x)dx = \int \frac{\cos x}{\sin x} dx$$

Make the substitution $u = \sin x$, then

$$\frac{du}{dx} = \cos x$$

and so

$$\int \frac{(\cos x)dx}{\sin x} = \int \frac{\dfrac{du}{dx}dx}{u} = \int \frac{du}{u} = \log_e u + c$$

Therefore

$$\int (\cot x)dx = \log_e(\sin x) + c \qquad\qquad (12.41)$$

EXERCISES 12

1. Convert the following angles in degrees to radians:

(a) $60°$ (b) $100°$ (c) $15°$

2. Convert the following angles in radians to degrees:

(a) 0.1 rad   (b) 2.7 rad   (c) 1.4 rad

3. Evaluate, using tables

(a) $\sin 65°$	(b) $\sec 42°$	(c) $\tan 70°$
(d) $\sin 130°$	(e) $\cos 110°$	(f) $\cot 140°$
(g) $\operatorname{cosec} 245°$	(h) $\cos 267°$	(i) $\tan 198°$
(j) $\operatorname{cosec} 314°$	(k) $\cos 300°$	(l) $\cot 343°$
(m) $\sin -198°$	(n) $\sec -285°$	(o) $\tan -100°$

4. Differentiate with respect to x:

(a) $x^2 \cdot \sin 2x$ (b) $x/(\tan x)$ (c) $\cos(2 - 3x)$

5. Find the co-ordinates of the maximum and minimum points on the curve of the function $y = 50e^{0.2 \cdot \sin x}$ in the range $0 \leqslant x \leqslant 2\pi$.

6. Find the following integrals, using the suggested substitutions:

(a) $\int(\sin 4x)dx : z = 4x$ (b) $\int x \cdot e^{-x^2} dx : z = x^2$

(c) $\int(\sin x \cdot \cos x)dx : z = \sin x$

7. Integrate by parts:

(a) $\int x \cdot \cos x \cdot dx$ (b) $\int e^x \sin x \cdot dx$

13

Differential Equations and their Solution

THE MICROBIAL CULTURE AGAIN

The idea of a differential equation was introduced in Chapter 6, and pages 103 to 110 should be re-read at this point before proceeding.

In the discussion of microbial culture growth, the aim was to obtain a mathematical function describing the increase of culture size with time, i.e. $n = f(t)$, by setting up a mathematical model of growth based on supposed features of microbial growth. The model considered two broad features of the dynamics of culture growth, and finally described the way that the rate of change of number of organisms was related to the number of organisms present in the culture. The relationship between rate of growth and organism number was a differential equation, 6.12, which was

$$\frac{dn}{dt} = kn \left(1 - \frac{n}{a} \right)$$

or, in a form more convenient for our present purpose,

$$\frac{dn}{dt} = \frac{kn}{a} (a - n) \tag{13.1}$$

This is a function of the form $dn/dt = f(n)$, but the function we require, $n = f(t)$, can be obtained by 'solving' the differential equation 13.1. To solve a differential equation, we evidently need to eliminate the differential coefficient which, of course, is done by integration.

Before proceeding to a more systematic study of differential equations and their solution, equation 13.1 will be solved, since, to anyone working systematically through the book, the equation must now be very familiar through constant use.

Although the core feature in the solution of a differential equation is integration, there is usually some preparatory work to be done first, and some 'tidying up' operations afterwards: so a number of distinct stages can usually be recognized. For 13.1, these stages are as follows.

(i) Re-arrange, to obtain terms in n and dn on the left-hand side, and dt on the right:

$$\frac{dn}{n(a-n)} = \frac{kdt}{a}.$$

(ii) Now, each side is integrated; the left-hand side with respect to n, and the right-hand side with respect to t. However, the left-hand side cannot be integrated as it stands, but must be split into partial fractions first (pages 129 to 133).
Hence

$$\frac{dn}{an} + \frac{dn}{a(a-n)} = \frac{kdt}{a}$$

and the a's cancel, giving

$$\frac{dn}{n} + \frac{dn}{a-n} = kdt \qquad (13.2)$$

If this is your first reading of the book, and you have missed out the section on 'integration by partial fractions' in Chapter 7, just accept the above result for the time being. It is, however, very easy to see that $1/n + 1/(a-n) = a/n(a-n)$: this is all we have done to the left-hand side of the equation. Now integrate 13.2:

$$\log_e n - \log_e(a-n) = kt + c$$

where c is the sum of the constants of integration of each of the separate integrals (separate terms) of 13.2.

(iii) Now it merely remains to write the answer in the most useful form. First, we have

$$\log_e\left(\frac{n}{a-n}\right) = kt + c$$

Next, take exponentials (anti-natural logarithms) of both sides:

$$\frac{n}{a-n} = e^{(kt+c)}$$

Now invert both sides, and split the right-hand side using the First Law of indices:

$$\frac{a-n}{n} = \frac{1}{e^{kt} \cdot e^c}$$

i.e.

$$\frac{a}{n} - 1 = e^{-kt} \cdot e^{-c}$$

Put $b = e^{-c}$, so that

$$\frac{a}{n} = 1 + be^{-kt}$$

Invert both sides again:

$$\frac{n}{a} = \frac{1}{1 + be^{-kt}}$$

and so

$$n = \frac{a}{1 + be^{-kt}} \qquad (13.3)$$

which is the logistic function (page 198). It is of course quite suitable for describing, at least approximately, the growth of a culture of micro-organisms.

Now that we have seen an example of the occurrence, use, and solution of a differential equation, we shall examine these equations in greater detail.

DEFINITIONS

Order

A differential equation may contain differential coefficients of any order, dy/dx, d^2y/dx^2, d^3y/dx^3, etc., and the order of a differential equation is the highest order differential coefficient contained in the equation. Hence

$$a\frac{d^2y}{dx^2} + b\frac{dy}{dx} + cy = 0$$

is a second-order differential equation.

Degree

The degree of a differential equation is the power of the highest order differential coefficient; thus

$$\left(\frac{d^2 y}{dx^2}\right)^3 - a\frac{dy}{dx} = c$$

is a second-order equation of the third degree, while

$$a\left(\frac{d^2 y}{dx^2}\right)^2 + b\left(\frac{dy}{dx}\right)^3 + cy - d = 0$$

is a second-order equation of the second degree, because the power of the highest order differential coefficient is 2.

General solution

When the solution of a differential equation contains a constant of integration, of unknown value, such as $b(= e^{-c})$ in equation 13.3, the solution is known as a general solution.

Particular solution

Sometimes, when additional information is available, the constant of integration can be evaluated. Then, the solution of the differential equation is a particular solution. For instance, let the initial inoculum in our microbial culture be n_0. Thus when $t = 0$, $n = n_0$. With this information, we can substitute in equation 13.3 for the unknown constant of integration:

$$n_0 = \frac{a}{1 + b}$$

i.e.

$$b = a/n_0 - 1$$

Thus 13.3 becomes

$$n = \frac{a}{1 + \left(\dfrac{a}{n_0} - 1\right) e^{-kt}} \tag{13.4}$$

and so 13.4 is a particular solution of the differential equation 13.1.

Fig. 13.1 shows the curves of three different logistic functions of the

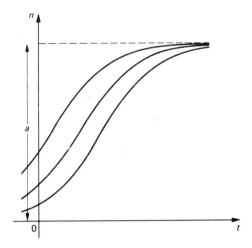

Fig. 13.1   Curves of the logistic function $n = a/(1 + be^{-kt})$ differing only in the value of b.

form of 13.3, in which the constants a and k are identical from one function (curve) to another. However, the constant b differs for each, and since $b = a/n_0 - 1$ this means that n_0 also differs. These functions could describe the growth of three different cultures of micro-organisms of the same species and under identical conditions; only the initial sizes of the inoculum, n_0, differ from one to another. On the other hand, the differential equation 13.1 describing the underlying dynamic situation which gives rise to the curves shown in Fig. 13.1 is the same for all three of the curves, since it only contains the constants a and k. This confirms that, under this model, the *dynamics* of culture growth are not affected by the size of the inoculum.

The general solution of equation 13.1 can be represented by an infinite number of curves of the form shown in Fig. 13.1, and each has the property that at any point the gradient is given by

$$\frac{dn}{dt} = kn\left(1 - \frac{n}{a}\right),$$

the differential equation with which we started. Any particular curve, such as one of those shown in Fig. 13.1 with a specified value of n_0 (and hence of b, and so in turn c, the constant of integration), represents a particular solution of the differential equation 13.1.

EQUATIONS WHERE THE VARIABLES ARE SEPARABLE

Equations of the form $dy/dx = f(x)$

These are very straightforward to solve, provided that $f(x)$ is integrable, since

$$y = \int f(x)\,dx$$

EXAMPLE 13.1
  Solve

$$\frac{dy}{dx} = bx$$

We have

$$y = b\int x\,dx$$

Hence

$$y = bx^2/2 + c.$$

Equations of the form $dy/dx = f(y)$

In these cases a cross-multiplication is first made to get all terms in y and dy on to the left-hand side, and dx on to the right-hand side of the equation, i.e.

$$\frac{dy}{f(y)} = dx$$

Any multiplying constant in $f(y)$ is best allowed to remain on the right-hand side. Then the left-hand side is integrated with respect to y, and the right-hand side with respect to x.

EXAMPLE 13.2
  Solve

$$\frac{dy}{dx} = by$$

Cross-multiply:

$$\frac{dy}{y} = b\,dx$$

Integrating,

$$\log_e y = bx + c$$

Take exponentials (i.e. anti-natural logarithms) of both sides,

$$y = e^{bx} \cdot e^c$$

Put $a = e^c$, then we have

$$y = ae^{bx}$$

which is the exponential function.

Equations of the form $f(y)dy = \phi(x)dx$

This is a special case of the form $dy/dx = F(x, y)$, where the differential coefficient is a function of both x and y. However, if we are to solve this equation, it must be capable of reduction to the form $f(y)dy = \phi(x)dx$, with all terms in y and dy on one side of the equation and all terms in x and dx on the other side.

EXAMPLE 13.3
Solve

$$\frac{dy}{dx} = \frac{(1+x)y}{x^2(1-y)}$$

Cross-multiply

$$\frac{(1-y)dy}{y} = \frac{(1+x)dx}{x^2}$$

i.e.

$$\frac{dy}{y} - dy = \frac{dx}{x^2} + \frac{dx}{x}$$

Integrating,

$$\log_e y - y = -\frac{1}{x} + \log_e x + c$$

This expression cannot be simplified any further: often we cannot set the solution of a differential equation in the form $y = f(x)$ explicitly. Nevertheless, the last line in this example is still the general solution of the differential equation, since the differential coefficient has been eliminated.

At this point, a first course in mathematics for the biologist ends; but you may dip into Chapter 14.

Cyclic phenomena
    Biology abounds with cyclic phenomena, which are mainly a resultant

of regular fluctuations of certain environmental factors, such as diurnal and seasonal rhythms. Such phenomena are readily described mathematically by using trigonometric functions.

EXAMPLE 13.4
  Solve

$$\frac{dy}{dt} = ky \cdot \cos t$$

Inspection of this equation shows that the variables are separable. The biologically interesting aspect of the equation, however, is that the rate of change of y with respect to t, dy/dt, is a function of both y and t. If this differential equation is describing the growth of a population, where y is the number of organisms present at time t, then the growth rate is dependent both upon actual number present, and upon time itself. As far as number of organisms present is concerned, dy/dt depends upon this factor in an exponential manner ($dy/dt = ky$); while population growth rate depends on time in a fluctuating manner. At zero time, the growth rate of the population is at its maximum (as far as time is concerned), since $\cos 0 = 1$. When $t = \pi/2$, growth rate is zero because $\cos \pi/2 = 0$ (see Fig. 12.6(b)). Hence the growth of the population is essentially exponential, though there is superimposed upon this a fluctuation due to time (perhaps seasonal). To solve the equation, first cross-multiply:

$$\frac{dy}{y} = k(\cos t)dt$$

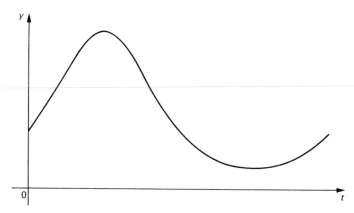

Fig. 13.2 The function $y = ae^{k \cdot \sin t}$.

Integrating,

$$\log_e \dot{y} = k \cdot \sin t + c$$

Taking exponentials of both sides gives

$$y = e^{k \cdot \sin t} \cdot e^c$$

and finally, putting $a = e^c$, we have

$$y = a e^{k \cdot \sin t}$$

Again, similarity to the exponential function is evident, and a graph of the function is shown in Fig. 13.2.

LINEAR DIFFERENTIAL EQUATIONS

If dy/dx and y are both present in a differential equation to the first degree only, then the equation is said to be linear. The most general form of a first order linear differential equation is

$$\frac{dy}{dx} + y \cdot f(x) = \phi(x) \tag{13.5}$$

where $f(x)$ and $\phi(x)$ are functions of x. Thus, for a differential equation to be linear, only dy/dx and y need be of first degree; the functions of x may take any form.

Equation 13.5 includes all the forms of the previous section. Thus, if $f(x) = 0$, $dy/dx = \phi(x)$; if $\phi(x) = 0$, $dy/dx = -y \cdot f(x)$, and the variables are separable. If neither of these cases apply, solution of linear equations is accomplished by first multiplying the whole equation by an 'integrating factor'. This puts the equation into a form in which it can be solved, provided that the resulting integral can be determined. Let us try to find the general form of the integrating factor for the linear equation 13.5. Multiply 13.5 throughout by v (the integrating factor), giving

$$v \frac{dy}{dx} + vy \cdot f(x) = v \cdot \phi(x) \tag{13.6}$$

Now

$$\frac{d(vy)}{dx} = v \frac{dy}{dx} + y \frac{dv}{dx}$$

(the rule for differentiating a product of two functions, equation (5.24)) i.e.

$$v \frac{dy}{dx} = \frac{d(vy)}{dx} - y \frac{dv}{dx} \, .$$

Substituting for $v \dfrac{dy}{dx}$ in 13.6 gives

$$\frac{d(vy)}{dx} - y \frac{dv}{dx} + vy \cdot f(x) = v \cdot \phi(x)$$

or

$$\frac{d(vy)}{dx} + y \left\{ v \cdot f(x) - \frac{dv}{dx} \right\} = v \cdot \phi(x) \tag{13.7}$$

Now select the value of v so that the coefficient of y in 13.7 is zero; so

$$v \cdot f(x) - \frac{dv}{dx} = 0$$

i.e.

$$\frac{dv}{v} = f(x)\,dx$$

Next, we integrate both sides, but as $f(x)$ is unspecified the right-hand side remains as an integral:

$$\log_e v = \int f(x)\,dx$$

the constant of integration may be assumed to be zero since this does not affect the method. So

$$v = e^{\int f(x)\,dx} \tag{13.8}$$

Substituting for v in equation 13.7, and remembering that this value of v implies that the coefficient of y is zero, gives

$$\frac{d(ye^{\int f(x)\,dx})}{dx} = \phi(x) \cdot e^{\int f(x)\,dx}$$

Integrate both sides with respect to x:

$$ye^{\int f(x)\,dx} = \int \{\phi(x) \cdot e^{\int f(x)\,dx}\}\,dx + c$$

i.e.

$$y = e^{-\int f(x)\,dx} \int \{\phi(x) \cdot e^{\int f(x)\,dx}\}\,dx + c \tag{13.9}$$

EXAMPLE 13.5
Solve

(a) $\dfrac{dy}{dx} + \dfrac{y}{x} = 6x^2$ (b) $\dfrac{dy}{dx} - y \cdot \tan x = \sin x$

(a) In terms of the general linear equation 13.5, $f(x) = 1/x$ and $\phi(x) = 6x^2$. The integrating factor is

$$e^{\int dx/x} = e^{\log_e x} = x$$

So, from 13.9 we have

$$y = \frac{1}{x} \int 6x^2 \cdot x \cdot dx = \frac{1}{x} \int 6x^3 \, dx$$

$$= \frac{1}{x} \left(\frac{6x^4}{4} + c \right)$$

i.e.

$$y = \frac{3x^3}{2} + \frac{c}{x}$$

(b) In terms of equation 13.5, $f(x) = -\tan x$ and $\phi(x) = \sin x$. The integrating factor is

$$e^{\int -(\tan x)dx} = e^{-\int (\tan x)dx} = e^{-\log_e(\sec x)} = \frac{1}{\sec x} = \cos x$$

Hence, from 13.9 we have

$$y = (\sec x) \int \sin x \cdot \cos x \cdot dx \qquad (13.10)$$

The integrand is a product of two functions of x, and so we integrate by parts. In relationship 7.19, putting $\psi(x) = \sin x$ and $\phi(x) = \cos x$, we have

$$\int \sin x \cdot \cos x \cdot dx = (\sin x) \int (\cos x)dx - \int \left\{ \frac{d(\sin x)}{dx} \int (\cos x)dx \right\} dx + c$$

i.e.

$$\int \sin x \cdot \cos x \cdot dx = \sin^2 x - \int \cos x \cdot \sin x \cdot dx + c$$

So

$$2 \int \sin x \cdot \cos x \cdot dx = \sin^2 x + c$$

i.e.

$$\int \sin x \cdot \cos x \cdot dx = \tfrac{1}{2}\sin^2 x + c$$

Substituting back into 13.10 gives

$$y = \sec x \cdot (\tfrac{1}{2} \sin^2 x + c)$$

The integral in 13.10 could equally well have been integrated by substitution.

Put $u = \sin x$, then $du/dx = \cos x$, and so $dx = du/\cos x$. Thus

$$\int \sin x \cdot \cos x \cdot dx = \int u \cdot \cos x \cdot \frac{du}{\cos x} = \int u\,du$$

$$= \frac{u^2}{2} + c = \frac{\sin^2 x}{2} + c$$

Substitution back into 13.10 gives the same answer as before.

EXERCISES 13

1. Solve (general solution) the following differential equations, in which the variables are separable.

(a) $\dfrac{dy}{dx} = 1 - \dfrac{y}{a}$ (b)* $x(x+1)\dfrac{dy}{dx} = y - 1$ (c)* $2x\dfrac{dy}{dx} = 1 - y^2$

(d)* $\sin y + e^x \cdot \dfrac{dy}{dx} = 0$

2.* Find the general solution to the following linear differential equations.

(a) $\dfrac{dy}{dx} + 2y = x$ (b) $\dfrac{dy}{dx} = 2 \cdot \cos 2x - y \cdot \cot x$

(c) $\cos x \cdot \dfrac{dy}{dx} + y \cdot \sin x = 1$

14

Matrix Algebra

MATRICES AND VECTORS

A matrix, plural **matrices**, is a collection of numbers (some or all of which may be algebraic symbols) arranged in a rectangular form with rows and columns, of which there may be any number. The whole matrix can be given a symbol of its own; e.g.

$$\mathbf{M} = \begin{bmatrix} 1 & -2 \\ 0 & 1 \end{bmatrix} \tag{14.1}$$

In relationship 14.1 we have a matrix consisting of two **rows** and two **columns** of figures and, to show that the four numbers must be regarded as an entity, they are enclosed in square brackets. The whole collection of numbers — the matrix — is given the symbol **M**, and matrices will always be denoted by symbols in heavy type. Hence **M** is distinguishable from M, the latter denoting a single number. The convention of enclosing the numbers within a matrix by square brackets is not universal; the use of rounded brackets is almost as common:

$$\mathbf{M} = \begin{pmatrix} 1 & -2 \\ 0 & 1 \end{pmatrix}$$

More unfortunate, however, is the fact that some authors do not use heavy type for symbolizing matrices; it is far more satisfactory to do so as ambiguities are thereby avoided.

The general matrix

The individual numbers within a matrix are called **elements**, and they are arranged in two dimensions — rows and columns. The **general matrix**, of which any particular matrix is a special case, has m rows and n columns. The usual way of writing the general matrix is to symbolize it by a capital letter in heavy type, in the usual way, and then to denote the individual elements by the corresponding ordinary lower case letter with two subscripts: the first subscript denotes the row to which the element belongs, and the second subscript the column. The general matrix can then be written out as shown below.

$$
\mathbf{A} =
\begin{bmatrix}
a_{11} & a_{12} & a_{13} \dots a_{1j} \dots a_{1n} \\
a_{21} & a_{22} & a_{23} \dots a_{2j} \dots a_{2n} \\
a_{31} & a_{32} & a_{33} \dots a_{3j} \dots a_{3n} \\
\vdots & \vdots & \vdots \qquad \vdots \qquad \vdots \\
a_{i1} & a_{i2} & a_{i3} \dots a_{ij} \dots a_{in} \\
\vdots & \vdots & \vdots \qquad \vdots \qquad \vdots \\
a_{m1} & a_{m2} & a_{m3} \dots a_{mj} \dots a_{mn}
\end{bmatrix}
\tag{14.2}
$$

Any particular element may be designated by its row number and its column number, that is by its subscripts. For example, the element of **A**, above, in the 4th row and the 2nd column is a_{42}, in the 3rd row and the 3rd column is a_{33}, and more generally, in the ith row and the jth column is a_{ij}. Notice that the row specification is *always* given first, and that the subscripts are not separated from one another by commas.

Matrix **A** in 14.2 has m rows and n columns; thus it is spoken of as an $(m \times n)$ matrix, and this is a convenient way of specifying the size of a matrix. The element a_{ij} is called the **general element**.

The transpose of a matrix

The **transpose** of a matrix is simply the matrix with its rows and columns interchanged, and is denoted by priming the capital letter designating the matrix. Thus, the transpose of **M** of 14.1 is given by

$$
\mathbf{M'} =
\begin{bmatrix}
1 & 0 \\
-2 & 1
\end{bmatrix}
$$

The symbol $\mathbf{M}^T$ is also used to denote a transposed matrix.

Particular types of matrix

Having introduced some points of terminology and notation, and also the general matrix as a framework of reference, we are now in a position to examine some important specific kinds of matrix.

Zero matrices
A **zero matrix** is one in which all elements are zero. It is denoted by **O**, and an example is

$$\mathbf{O} = \begin{bmatrix} 0 & 0 & 0 & 0 \\ 0 & 0 & 0 & 0 \\ 0 & 0 & 0 & 0 \end{bmatrix}$$

Square matrices
A **square matrix** is one in which the numbers of rows and columns are equal, and such matrices are of outstanding importance. The size can be indicated either by specifying that the matrix is an $(m \times m)$ one, or that the matrix is of **order** m. Matrix **M** of 14.1 is a square matrix of order 2. Matrix **A** of 14.3 below is a (4×4) matrix, or alternatively a square matrix of order 4.

$$\mathbf{A} = \begin{bmatrix} 2 & -1 & -4 & 5 \\ 2 & 1 & -3 & 2 \\ 2 & 3 & 0 & 2 \\ 2 & 5 & 5 & 7 \end{bmatrix} \qquad (14.3)$$

There are two categories of **diagonal elements** in a square matrix; one from top-left to bottom-right, and the other from top-right to bottom-left. The former diagonal is particularly important, and is known as the **leading diagonal**; while the diagonal from top-right to bottom-left has no significance at all.

Symmetric matrices
A **symmetric matrix** is a square matrix which is identical with its transpose. Thus

$$\mathbf{S} = \begin{bmatrix} 3 & -1 & 2 \\ -1 & 0 & 1 \\ 2 & 1 & 4 \end{bmatrix} \qquad (14.4)$$

is a symmetric matrix, as it is unchanged when rows and colums are inter-changed. The name arises because the values of the elements are symmetri-cally disposed about the leading diagonal; it can be said that the elements on one side of the leading diagonal are 'reflected' on the other side, and the leading diagonal acts as though it were a plane mirror. Because of this, a symmetric matrix need not be quoted in full, but in one of two partial ways. Thus matrix **S** of 14.4 can be written either as

$$\begin{bmatrix} 3 & -1 & 2 \\ & 0 & 1 \\ & & 4 \end{bmatrix} \quad \text{or as} \quad \begin{bmatrix} 3 & & \\ -1 & 0 & \\ 2 & 1 & 4 \end{bmatrix}$$

Diagonal matrices

A **diagonal matrix** is a square matrix in which all elements not on the leading diagonal are zero. Thus

$$\mathbf{D} = \begin{bmatrix} 2 & 0 & 0 \\ 0 & -1 & 0 \\ 0 & 0 & -5 \end{bmatrix} \tag{14.5}$$

is a diagonal matrix. Such matrices are particularly simple to deal with.

Unit matrices

The **unit matrix** is a diagonal matrix in which all elements of the leading diagonal are unity. The unit matrix is always denoted by **I** or as $\mathbf{I_n}$, where n is the order of the matrix; for example

$$\mathbf{I_4} = \begin{bmatrix} 1 & 0 & 0 & 0 \\ 0 & 1 & 0 & 0 \\ 0 & 0 & 1 & 0 \\ 0 & 0 & 0 & 1 \end{bmatrix} \tag{14.6}$$

is the 4th order unit matrix. Unit matrices have interesting properties, and play a leading role in matrix algebra.

Column vectors

If in the general matrix $n = 1$, then the resulting matrix has only one column with m rows. A logical name would be 'column matrix', but it is more usually known as a **column vector** for a reason that will be explained later (page 261). A column vector is normally denoted by a small (lower case) letter in heavy type, e.g.

$$\mathbf{a} = \begin{bmatrix} a_{11} \\ a_{21} \\ a_{31} \\ \vdots \\ a_{i1} \\ \vdots \\ a_{m1} \end{bmatrix} \tag{14.7}$$

Row vectors

Referring again to the general matrix, if $m = 1$ then the resulting matrix has only one row with n columns, and is known as a **row vector**. In theory and in applications, the column vector is the more fundamental, and so a row vector is usually regarded as the transpose of a column vector and is denoted by a primed small letter in heavy type:

$$\mathbf{a}' = [a_{11} \quad a_{12} \quad a_{13} \ldots a_{1j} \ldots a_{1n}] \tag{14.8}$$

The constant subscript 1 may be omitted in a row vector, the elements being labelled a_1 to a_n; and similarly in a column vector the elements will be labelled a_1 to a_m.

Geometrical interpretation of a column vector

If you have pursued a course in mechanics or elementary physics, you may remember that a **vector** is a quantity having both magnitude and direction, as opposed to a **scalar** quantity which has magnitude only. Thus, ordinary numbers are scalar quantities, but a force, which has a magnitude and is applied in a particular direction, is a vector quantity. Row and column vectors are not, however, called vectors because they are orientated horizontally and vertically, respectively, but for an entirely different reason.

Consider the column vector $\mathbf{x} = \begin{bmatrix} 2 \\ 1 \end{bmatrix}$. Now the two numbers within the column vector can be thought of as the co-ordinates of a point P situated in a plane defined by the usual $x–y$ axes (Fig. 14.1). But the column vector is not merely another way of specifying the co-ordinates of a point; it means something more than that. Join the point, whose co-ordinates are given by the elements of the column vector, to the origin by a straight

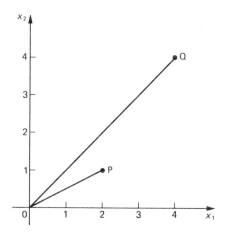

Fig. 14.1   Vectors OP and OQ (see page 261).

line. This line is a vector quantity; it has both magnitude (i.e. length), given by $\sqrt{\{(2)^2 + (1)^2\}} = \sqrt{5}$ (see page 32), and direction or orientation, given by $\tan^{-1}(1/2) \simeq 26°34'$ (i.e. the angle whose tangent is $1/2$). The angle is always measured in an anti-clockwise direction from the horizontal axis. Similarly, another vector $y = \begin{bmatrix} 4 \\ 4 \end{bmatrix}$ represents the straight line OQ with a length of $\sqrt{\{(4)^2 + (4)^2\}} = \sqrt{32}$, and an orientation of $\tan^{-1}(4/4) = 45°$.

Thus we see that a two-component column vector can represent a straight line in two-dimensional space from the origin to the point whose co-ordinates are specified by the elements of the column vector. By analogy, a three-component column vector may represent a straight line in three dimensional space, again from the origin to the point whose components are specified by the elements of the vector. The analogy may be pursued further, and we conclude that an n-component column vector can represent a similar straight line in a space of n dimensions; it cannot be illustrated, but it can be described mathematically.

ELEMENTS OF MATRIX ALGEBRA

Since any kind of matrix can be given a unique symbol, this means that the symbol represents the whole array of numbers within the matrix. If we have two different matrices, each represented by a single symbol, one might be able to operate on these symbols as can be done with symbols representing single numbers in ordinary algebra. For instance, let **A** and **B** denote two matrices; then one *could* write such expressions as **A** + **B**, and **A** · **B**. If such operations could be performed, then presumably the results would be other matrices, i.e.

$$A + B = C$$

and

$$A \cdot B = D$$

$$(14.9)$$

Hence, three questions may be posed. (1) Are operations such as those represented by the equations in 14.9 possible? (2). If so, under what conditions are they possible? (3) What are the new matrices, e.g. **C** and **D**, which are formed by such operations? The answer to Question 1 is mostly affirmitive, and in considering the various facets of matrix algebra, the answers to Questions 2 and 3 will emerge.

Matrix equality

Two matrices are equal to one another only if each contains the same number of rows and the same number of columns, and if *all* corresponding elements are equal to one another. For the following two matrices, **A** and **B**, we have

$$\begin{bmatrix} a_{11} & a_{12} & a_{13} \\ a_{21} & a_{22} & a_{23} \end{bmatrix} = \begin{bmatrix} b_{11} & b_{12} & b_{13} \\ b_{21} & b_{22} & b_{23} \end{bmatrix} \qquad (14.10)$$

$$A = B$$

only if $a_{11} = b_{11}, a_{21} = b_{21}, \ldots, a_{23} = b_{23}$. To put the condition more briefly, we write

$$a_{ij} = b_{ij} \quad \text{for all} \quad i, j.$$

Addition and subtraction

The operations of addition and subtraction can be performed and, being very similar, they can be considered together. Two matrices can be added to, or subtracted from, one another only if each contains the same number of rows and the same number of columns; and the operations are performed simply by adding or subtracting corresponding elements.

EXAMPLE 14.1

If
$$A = \begin{bmatrix} 4 & 6 \\ 5 & 7 \end{bmatrix} \quad \text{and} \quad B = \begin{bmatrix} 2 & 3 \\ 1 & 4 \end{bmatrix}$$

Find

(a) $A + B$ (b) $B + A$ (c) $A - B$ (d) $B - A$

(a)
$$\begin{bmatrix} 4 & 6 \\ 5 & 7 \end{bmatrix} + \begin{bmatrix} 2 & 3 \\ 1 & 4 \end{bmatrix} = \begin{bmatrix} (4+2) & (6+3) \\ (5+1) & (7+4) \end{bmatrix} = \begin{bmatrix} 6 & 9 \\ 6 & 11 \end{bmatrix}$$

(b) Proceeding in the same way as above:
$$\begin{bmatrix} 2 & 3 \\ 1 & 4 \end{bmatrix} + \begin{bmatrix} 4 & 6 \\ 5 & 7 \end{bmatrix} = \begin{bmatrix} 6 & 9 \\ 6 & 11 \end{bmatrix}$$

(c)
$$\begin{bmatrix} 4 & 6 \\ 5 & 7 \end{bmatrix} - \begin{bmatrix} 2 & 3 \\ 1 & 4 \end{bmatrix} = \begin{bmatrix} (4-2) & (6-3) \\ (5-1) & (7-4) \end{bmatrix} = \begin{bmatrix} 2 & 3 \\ 4 & 3 \end{bmatrix}$$

(d)
$$\begin{bmatrix} 2 & 3 \\ 1 & 4 \end{bmatrix} - \begin{bmatrix} 4 & 6 \\ 5 & 7 \end{bmatrix} = \begin{bmatrix} (2-4) & (3-6) \\ (1-5) & (4-7) \end{bmatrix} = \begin{bmatrix} -2 & -3 \\ -4 & -3 \end{bmatrix} = -\begin{bmatrix} 2 & 3 \\ 4 & 3 \end{bmatrix}$$

This example shows that:

(i) $A + B = B + A$

(ii) $A - B = -(B - A)$;

that is, addition and subtraction of matrices are commutative processes, as is the case for single quantities in ordinary algebra. In other words, whenever two matrices are added together, the sum is the same regardless of which matrix appears first; similarly, if one matrix is subtracted from another, the same *numerical* result is obtained whichever way round the subtraction is performed, but the sign of the result depends on the order of the two given matrices.

Care must be taken when the elements in the matrices differ in sign, as in the next example.

EXAMPLE 14.2

If
$$P = \begin{bmatrix} 1 & 2 & -1 \\ -3 & -2 & 1 \\ 2 & 3 & 2 \end{bmatrix} \quad \text{and} \quad Q = \begin{bmatrix} 3 & 2 & -1 \\ 4 & 2 & -2 \\ -1 & 3 & 1 \end{bmatrix},$$

Find

(a) **P** + **Q**   (b) **P** − **Q**

(a)

$$
\begin{bmatrix} 1 & 2 & -1 \\ -3 & -2 & 1 \\ 2 & 3 & 2 \end{bmatrix} + \begin{bmatrix} 3 & 2 & -1 \\ 4 & 2 & -2 \\ -1 & 3 & 1 \end{bmatrix}
$$

$$
= \begin{bmatrix} \{1+3\} & \{2+2\} & \{(-1)+(-1)\} \\ \{(-3)+4\} & \{(-2)+2\} & \{1+(-2)\} \\ \{2+(-1)\} & \{3+3\} & \{2+1\} \end{bmatrix}
$$

$$
= \begin{bmatrix} 4 & 4 & -2 \\ 1 & 0 & -1 \\ 1 & 6 & 3 \end{bmatrix}
$$

(b)

$$
\begin{bmatrix} 1 & 2 & -1 \\ -3 & -2 & 1 \\ 2 & 3 & 2 \end{bmatrix} - \begin{bmatrix} 3 & 2 & -1 \\ 4 & 2 & -2 \\ -1 & 3 & 1 \end{bmatrix}
$$

$$
= \begin{bmatrix} \{1-3\} & \{2-2\} & \{(-1)-(-1)\} \\ \{(-3)-4\} & \{(-2)-2\} & \{1-(-2)\} \\ \{2-(-1)\} & \{3-3\} & \{2-1\} \end{bmatrix}
$$

$$
= \begin{bmatrix} -2 & 0 & 0 \\ -7 & -4 & 3 \\ 3 & 0 & 1 \end{bmatrix}
$$

The operation of addition (or subtraction) may be summarized for two general matrices. Let

$$
\mathbf{A} = \begin{bmatrix} a_{11} & a_{12} & \cdots a_{1n} \\ a_{21} & a_{22} & \cdots a_{2n} \\ \vdots & \vdots & \vdots \\ a_{m1} & a_{m2} & \cdots a_{mn} \end{bmatrix} \quad \text{and} \quad \mathbf{B} = \begin{bmatrix} b_{11} & b_{12} & \cdots b_{1n} \\ b_{21} & b_{22} & \cdots b_{2n} \\ \vdots & \vdots & \vdots \\ b_{m1} & b_{m2} & \cdots b_{mn} \end{bmatrix}
$$

Define a matrix $\mathbf{C}$ such that $\mathbf{A} + \mathbf{B} = \mathbf{C}$, then

$$\mathbf{C} = \begin{bmatrix} (a_{11} + b_{11}) & (a_{12} + b_{12}) & \ldots (a_{1n} + b_{1n}) \\ (a_{21} + b_{21}) & (a_{22} + b_{22}) & \ldots (a_{2n} + b_{2n}) \\ \vdots & \vdots & \vdots \\ (a_{m1} + b_{m1}) & (a_{m2} + b_{m2}) & \ldots (a_{mn} + b_{mn}) \end{bmatrix} \quad (14.11)$$

Subtraction proceeds in an identical way, with the plus signs replaced by minus signs.

Multiplication

Multiplication of a matrix by a scalar

The word 'scalar' here refers to a single number which is not an element of a matrix. It is possible to multiply a matrix by a scalar quantity. Suppose that matrix $\mathbf{A}$ is given by

$$\mathbf{A} = \begin{bmatrix} a_{11} & a_{12} \\ a_{21} & a_{22} \end{bmatrix}$$

then if k is a scalar number, we have

$$k\mathbf{A} = k \begin{bmatrix} a_{11} & a_{12} \\ a_{21} & a_{22} \end{bmatrix} = \begin{bmatrix} ka_{11} & ka_{12} \\ ka_{21} & ka_{22} \end{bmatrix} \quad (14.12)$$

So, multiplying a matrix by a scalar simply implies that every element in the matrix is multiplied by the scalar.

Multiplication of two matrices

The multiplication of two matrices is a feasible operation, but note at once that it is not done in a similar way to addition and subtraction; that is, one does not merely multiply together corresponding elements in the two matrices.

You will perhaps conclude from this that matrix algebra is an illogical subject. Before considering the details of matrix multiplication, let us look a little further into the nature of ordinary algebra performed with single quantities, and make some comparisons with matrix algebra.

Ordinary algebra is a 'natural' process in that its rules derive ultimately from the fundamental process of addition. Two added to three must equal five, thus giving the inescapable result that $3 + 2 = 5$. Subtraction arises naturally from addition by invoking the idea of a negative number and using it in the process of addition in the usual way; thus $3 + (-2) = 3 - 2 = 1$.

Multiplication is continued addition. Thus, 3 x 2 is short for 3 + 3, or 2 + 2 + 2, which equals 6; and 3 x 5 is an abbreviation for 3 + 3 + 3 + 3 + 3, or 5 + 5 + 5, which is 15. Division is continued subtraction. To see this, consider 24 ÷ 8 = 3. Starting with 24, subtract 8 leaving 16; subtract 8 again leaving 8, and subtract 8 once more which yields 0. Three subtractions of 8 from 24 give zero, i.e. 24 − 8 − 8 − 8 = 0 or 24 − (8 + 8 + 8) = 0. This is the same as writing 24 − (3 x 8) = 0. If the divisor does not go into the dividend exactly, the actual demonstration that division is continued subtraction is not quite as straightforward, but the same principle applies. Hence, as division is continued subtraction, and subtraction is merely addition involving negative numbers, we see that the four basic rules of ordinary algebra are derived from the fundamental process of addition.

Since a matrix is an entity consisting of more than one number, there is clearly no fundamental or even single way in which one matrix can be combined with another in an algebraic operation. Hence the rules in matrix algebra are not natural ones, but are *conventions*, and they have been formulated in the way they have in order to make the fullest use of being able to apply algebraic processes to whole arrays. At first sight, the condition and rule for matrix multiplication may seem very complicated and highly artificial, but practice will show that the rule is not difficult to master, and will also demonstrate the great utility of the procedure.

First of all, two matrices can be multiplied together only if the number of columns in the first matrix is equal to the number of rows in the second. For example, let

$$A = \begin{bmatrix} 1 & -2 \\ 3 & 2 \end{bmatrix} \quad B = \begin{bmatrix} 4 & 0 & 2 \\ -1 & 6 & 3 \end{bmatrix}$$

Then the multiplication $A \cdot B$ is possible, as the number of columns in $A (= 2)$ is equal to the number of rows in $B (= 2)$. However, there can be no product $B \cdot A$ because the number of columns in $B (= 3)$ is not equal to the number of rows in $A (= 2)$. So although a given pair of matrices can be multiplied together one way round, the operation cannot necessarily be performed with the order of the matrices reversed. This situation, which is a consequence of the rule for the multiplication of matrices, is clearly very different from the situation involving single quantities in ordinary algebra. In the latter case, for two numbers a and b, $ab = ba$; multiplication is commutative. However, for two matrices, A and B, although $A \cdot B$ may be feasible, $B \cdot A$ may be impossible. Thus, multiplication of matrices is not, in general, commutative. Even though in a particular instance both the operations $A \cdot B$ and $B \cdot A$ are possible, the resultant matrices will usually be different (Example 14.4(a) and (b)).

Returning to the two matrices introduced above, let us form the matrix product $\mathbf{A} \cdot \mathbf{B}$. The rule states that we must combine the rows of the first matrix with the columns of the second, in the following way:

$$\begin{bmatrix} 1 & -2 \\ 3 & 2 \end{bmatrix} \begin{bmatrix} 4 & 0 & 2 \\ -1 & 6 & 3 \end{bmatrix}$$

$$= \begin{bmatrix} \{(1)(4) + (-2)(-1)\} & \{(1)(0) + (-2)(6)\} & \{(1)(2) + (-2)(3)\} \\ \{(3)(4) + (2)(-1)\} & \{(3)(0) + (2)(6)\} & \{(3)(2) + (2)(3)\} \end{bmatrix}$$

$$= \begin{bmatrix} \{4 + 2\} & \{0 + (-12)\} & \{2 + (-6)\} \\ \{12 + (-2)\} & \{0 + 12\} & \{6 + 6\} \end{bmatrix} = \begin{bmatrix} 6 & -12 & -4 \\ 10 & 12 & 12 \end{bmatrix}$$

Notice that we have multiplied a (2 x 2) matrix by a (2 x 3) one, and obtained a (2 x 3) product matrix: we could write *Easy by definition*

$$(2 \times 2)(2 \times 3) = (2 \times 3) \qquad c = ? \qquad (14.13)$$

for describing the sizes of the matrices involved in the multiplication. Since the first number in any of the above brackets describing matrix size refers to the number of rows, and the second number in the brackets refers to the number of columns; and because multiplication requires the number of columns in the first matrix to be equal to the number of rows in the second, the two 'inner' numbers in the bracket combination on the left-hand side of 14.13, i.e. (2 x 2)(2 x 3), must be the same. Moreover, the dimensions of the product matrix are given by the two 'outer' numbers in this bracket combination, (2 x 3) in this example.

In general, for two matrices, $\mathbf{A}$ and $\mathbf{B}$, of sizes $(m_A \times n_A)$ and $(m_B \times n_B)$, respectively, the product $\mathbf{A} \cdot \mathbf{B}$ can be formed only if $n_A = m_B$, and the resulting product matrix is of size $(m_A \times n_B)$.

To demonstrate the multiplication rule for two general matrices, as was done for addition in 14.11, would be very cumbersome, so we shall show the rule in symbol notation for the two small matrices shown below. Let

$$\mathbf{A} = \begin{bmatrix} a_{11} & a_{12} & a_{13} \\ a_{21} & a_{22} & a_{23} \end{bmatrix} \quad \mathbf{B} = \begin{bmatrix} b_{11} & b_{12} \\ b_{21} & b_{22} \\ b_{31} & b_{32} \end{bmatrix}$$

Define a matrix $\mathbf{C}$ such that $\mathbf{A} \cdot \mathbf{B} = \mathbf{C}$, then   (14.14)

$$\mathbf{C} = \begin{bmatrix} (a_{11}b_{11} + a_{12}b_{21} + a_{13}b_{31}) & (a_{11}b_{12} + a_{12}b_{22} + a_{13}b_{32}) \\ (a_{21}b_{11} + a_{22}b_{21} + a_{23}b_{31}) & (a_{21}b_{12} + a_{22}b_{22} + a_{23}b_{32}) \end{bmatrix}$$

Notice that $\mathbf{A}$ is of size (2 x 3) and $\mathbf{B}$ is of size (3 x 2), thus $n_A = m_B = 3$, and the size of the product matrix is $(m_A \times n_B) = (2 \times 2)$.

EXAMPLE 14.3
   Define

$$A = \begin{bmatrix} 4 & -6 \\ 2 & 5 \end{bmatrix} \qquad B = \begin{bmatrix} 1 & 4 \\ 2 & -2 \\ 3 & -1 \end{bmatrix} \qquad c = \begin{bmatrix} -4 \\ 6 \end{bmatrix}$$

Find, if possible,

(a) $A \cdot B$ (b) $B \cdot A$ (c) $A \cdot c$
(d) $c \cdot A$ (e) $B \cdot c$ (f) $c \cdot B$

(a) A is a (2 x 2) matrix, and B is a (3 x 2) one; and the potential product can be represented as (2 x 2)(3 x 2). So the number of columns in the first matrix is not equal to the number of rows in the second, and thus the product $A \cdot B$ cannot be formed.
(b) This product is possible, and we have

$$\begin{bmatrix} 1 & 4 \\ 2 & -2 \\ 3 & -1 \end{bmatrix} \begin{bmatrix} 4 & -6 \\ 2 & 5 \end{bmatrix} = \begin{bmatrix} (4+8) & (-6+20) \\ (8-4) & (-12-10) \\ (12-2) & (-18-5) \end{bmatrix} = \begin{bmatrix} 12 & 14 \\ 4 & -22 \\ 10 & -23 \end{bmatrix}$$

(c) This product is also feasible:

$$\begin{bmatrix} 4 & -6 \\ 2 & 5 \end{bmatrix} \begin{bmatrix} -4 \\ 6 \end{bmatrix} = \begin{bmatrix} (-16-36) \\ (-8+30) \end{bmatrix} = \begin{bmatrix} -52 \\ 22 \end{bmatrix}$$

(d) This product cannot be formed.
(e) This product can be formed:

$$\begin{bmatrix} 1 & 4 \\ 2 & -2 \\ 3 & -1 \end{bmatrix} \begin{bmatrix} -4 \\ 6 \end{bmatrix} = \begin{bmatrix} (-4+24) \\ (-8-12) \\ (-12-6) \end{bmatrix} = \begin{bmatrix} 20 \\ -20 \\ -18 \end{bmatrix}$$

(f) This product is impossible.

When two matrices can be multiplied together in the order given, they are said to be conformable for multiplication. Thus, in Example 14.3, $B \cdot A$ is conformable, whereas $A \cdot B$ is not conformable for multiplication.
   Another point of terminology arises because the order in which the matrices are taken for multiplication is important. In Example 14.3(b), A is said to be **pre-multiplied by** B, or one could equally say that B is **post-multiplied** by A.

EXAMPLE 14.4

Let

$$A = \begin{bmatrix} -1 & 1 \\ 2 & -3 \end{bmatrix} \quad B = \begin{bmatrix} 2 & 3 \\ -4 & 5 \end{bmatrix}$$

$$I = \begin{bmatrix} 1 & 0 \\ 0 & 1 \end{bmatrix} \quad O = \begin{bmatrix} 0 & 0 \\ 0 & 0 \end{bmatrix}$$

Find

(a) **A · B**   (b) **B · A**   (c) **A · I**
(d) **I · A**   (e) **B · O**   (f) **O · B**

All these products are feasible: why?

(a)

$$\begin{bmatrix} -1 & 1 \\ 2 & -3 \end{bmatrix} \begin{bmatrix} 2 & 3 \\ -4 & 5 \end{bmatrix} = \begin{bmatrix} (-2-4) & (-3+5) \\ (4+12) & (6-15) \end{bmatrix} = \begin{bmatrix} -6 & 2 \\ 16 & -9 \end{bmatrix}$$

(b)

$$\begin{bmatrix} 2 & 3 \\ -4 & 5 \end{bmatrix} \begin{bmatrix} -1 & 1 \\ 2 & -3 \end{bmatrix} = \begin{bmatrix} (-2+6) & (2-9) \\ (4+10) & (-4-15) \end{bmatrix} = \begin{bmatrix} 4 & -7 \\ 14 & -19 \end{bmatrix}$$

(c)

$$\begin{bmatrix} -1 & 1 \\ 2 & -3 \end{bmatrix} \begin{bmatrix} 1 & 0 \\ 0 & 1 \end{bmatrix} = \begin{bmatrix} (-1+0) & (0+1) \\ (2-0) & (0-3) \end{bmatrix} = \begin{bmatrix} -1 & 1 \\ 2 & -3 \end{bmatrix}$$

(d)

$$\begin{bmatrix} 1 & 0 \\ 0 & 1 \end{bmatrix} \begin{bmatrix} -1 & 1 \\ 2 & -3 \end{bmatrix} = \begin{bmatrix} (-1+0) & (1+0) \\ (0+2) & (0-3) \end{bmatrix} = \begin{bmatrix} -1 & 1 \\ 2 & -3 \end{bmatrix}$$

(e)

$$\begin{bmatrix} 2 & 3 \\ -4 & 5 \end{bmatrix} \begin{bmatrix} 0 & 0 \\ 0 & 0 \end{bmatrix} = \begin{bmatrix} 0 & 0 \\ 0 & 0 \end{bmatrix}$$

(f)

$$\begin{bmatrix} 0 & 0 \\ 0 & 0 \end{bmatrix} \begin{bmatrix} 2 & 3 \\ -4 & 5 \end{bmatrix} = \begin{bmatrix} 0 & 0 \\ 0 & 0 \end{bmatrix}$$

In Example 14.4: (a) and (b) show that even if two matrices can be multiplied both ways, the answer is different in each case. This emphasises that, in general, matrix multiplication is not commutative.

In (c) and (d) it is shown that the unit matrix, **I**, behaves in matrix algebra in a similar manner to the number 1 in ordinary algebra. A matrix, **A**, can be pre- or post-multiplied by **I**, and the result will always be **A**; so that multiplication involving **I** *is* commutative, i.e.

$$\mathbf{A} \cdot \mathbf{I} = \mathbf{I} \cdot \mathbf{A} = \mathbf{A} \qquad (14.15)$$

A similar situation exists for the zero matrix in (e) and (f). This matrix behaves analagously to zero of ordinary algebra. Any matrix, **A**, pre- or post-multiplied by **O** yields the zero matrix; so that, again, multiplication involving **O** is commutative, i.e.

$$\mathbf{A} \cdot \mathbf{O} = \mathbf{O} \cdot \mathbf{A} = \mathbf{O} \qquad (14.16)$$

The reader may now like to examine some biological applications on pages 284 to 288, most of which can be read at this stage.

Inversion

Direct division of matrices is not defined, but under certain conditions, one matrix can be 'divided' into another by an indirect process involving multiplication. The procedure is best demonstrated in relation to ordinary numbers.

Suppose we require $12 \div 4$. This could be evaluated as 12×0.25, i.e. $12 \times 1/4$ or 12×4^{-1}. The number 4^{-1} is the reciprocal, or **inverse**, of 4, and the expression $12 \times 4^{-1} = 12 \div 4 = 3$. The inverse of a number is such that when it is multiplied by the number itself, the product is unity; e.g. $4 \times 4^{-1} = 4 \times 1/4 = 1$. In general terms, we have

$$x \cdot x^{-1} = 1 \qquad (14.17)$$

A square matrix can usually (but not always) be inverted, which means that when the inverse of the matrix is multiplied by the matrix itself then the product is the unit matrix, which we have already seen has similar properties in matrix algebra to the number 1 in ordinary algebra. So we have that

$$\mathbf{A} \cdot \mathbf{A}^{-1} = \mathbf{I} \qquad (14.18)$$

where **A** is a square matrix, and $\mathbf{A}^{-1}$ is the corresponding inverse matrix.

Inverse matrices are very important in matrix algebra, and this section has served to introduce the concept. The process for inverting a matrix is not straightforward, however, and other facets of matrix algebra need to be explained before the procedure of matrix inversion can be described.

DETERMINANTS

The inverse, and determinant, of a second order matrix

It has already been stated that the inverse, $\mathbf{A}^{-1}$, of any square matrix $\mathbf{A}$, is such that $\mathbf{A} \cdot \mathbf{A}^{-1} = \mathbf{I}$. Consider the second order matrix

$$\begin{bmatrix} a & c \\ b & d \end{bmatrix}$$

where, for clarity, we use a separate letter for each element rather than subscript notation.

Now, form a new matrix from the above, by reversing the two elements in the leading diagonal and changing the signs of the other two; i.e.

$$\begin{bmatrix} d & -c \\ -b & a \end{bmatrix}$$

and consider the product of these two matrices:

$$\begin{bmatrix} a & c \\ b & d \end{bmatrix}\begin{bmatrix} d & -c \\ -b & a \end{bmatrix} = \begin{bmatrix} (ad-bc) & (-ac+ac) \\ (bd-bd) & (-bc+ad) \end{bmatrix} = \begin{bmatrix} ad-bc & 0 \\ 0 & ad-bc \end{bmatrix}$$

$$= (ad-bc)\begin{bmatrix} 1 & 0 \\ 0 & 1 \end{bmatrix} = (ad-bc)\mathbf{I}$$

The product of these two matrices yields the unit matrix multiplied by the scalar quantity, $(ad-bc)$. Now consider the following matrix multiplication:—

$$\begin{bmatrix} a & c \\ b & d \end{bmatrix}\begin{bmatrix} \dfrac{d}{ad-bc} & \dfrac{-c}{ad-bc} \\ \dfrac{-b}{ad-bc} & \dfrac{a}{ad-bc} \end{bmatrix} = \begin{bmatrix} \dfrac{ad}{ad-bc}+\dfrac{-bc}{ad-bc} & \dfrac{-ac}{ad-bc}+\dfrac{ac}{ad-bc} \\ \dfrac{bd}{ad-bc}+\dfrac{-bd}{ad-bc} & \dfrac{-bc}{ad-bc}+\dfrac{ad}{ad-bc} \end{bmatrix}$$

$$\begin{bmatrix} \dfrac{ad-bc}{ad-bc} & 0 \\ 0 & \dfrac{ad-bc}{ad-bc} \end{bmatrix} = \begin{bmatrix} 1 & 0 \\ 0 & 1 \end{bmatrix}$$

This time, the product is the unit matrix, and since the matrix

$$\begin{bmatrix} \dfrac{d}{ad-bc} & \dfrac{-c}{ad-bc} \\ \dfrac{-b}{ad-bc} & \dfrac{a}{ad-bc} \end{bmatrix} \quad \text{has been formed entirely from} \quad \begin{bmatrix} a & c \\ b & d \end{bmatrix}$$

then the former matrix must be the inverse of the latter. The quantity $(ad - bc)$ is called the **determinant** of the matrix and, as already mentioned, is a scalar quantity.

The rule for forming the inverse of a second order matrix can therefore be put into words as follows. **The inverse of a second order matrix is formed by reversing the elements in the leading diagonal, changing the signs of the remaining two elements, and then dividing each of the resulting new elements by the determinant of the matrix.**

EXAMPLE 14.5

Find the inverse of

$$\begin{bmatrix} 4 & -6 \\ 2 & 5 \end{bmatrix}$$

and verify the result.

The determinant is $(4)(5) - (2)(-6) = 20 - (-12) = 32$. Thus the inverse is

$$\frac{1}{32} \begin{bmatrix} 5 & 6 \\ -2 & 4 \end{bmatrix} = \begin{bmatrix} \dfrac{5}{32} & \dfrac{3}{16} \\ \dfrac{-1}{16} & \dfrac{1}{8} \end{bmatrix}$$

Verifying:

$$\frac{1}{32} \begin{bmatrix} 4 & -6 \\ 2 & 5 \end{bmatrix} \begin{bmatrix} 5 & 6 \\ -2 & 4 \end{bmatrix} = \frac{1}{32} \begin{bmatrix} (20+12) & (24-24) \\ (10-10) & (12+20) \end{bmatrix}$$

$$= \frac{1}{32} \begin{bmatrix} 32 & 0 \\ 0 & 32 \end{bmatrix} = \begin{bmatrix} 1 & 0 \\ 0 & 1 \end{bmatrix}$$

Determinants arose independently of, and long before, matrix algebra. In order to introduce the notation of the subject, consider the matrix

$$A = \begin{bmatrix} a & c \\ b & d \end{bmatrix}$$

The determinant of **A** can be symbolized in various ways:

$$|A| = \det \mathbf{A} = \Delta = \begin{vmatrix} a & c \\ b & d \end{vmatrix} = ad - bc \qquad (14.19)$$

The first symbol in 14.19 shows the letter denoting the matrix, not in heavy type, but enclosed by vertical lines; the second symbol is obvious, while the third symbol is ambiguous since it can refer to the determinant of any matrix. The fourth symbol looks like the original matrix, but vertical lines replace the square brackets; and finally in 14.19 is the definition of the determinant.

Third and higher order determinants

The idea of a determinant can be extended to higher orders. Consider the 3 x 3 array

$$\mathbf{A} = \begin{bmatrix} a & d & g \\ b & e & h \\ c & f & i \end{bmatrix}$$

where again for clarity the subscript notation is not used. What meaning is to be ascribed to $|A|$?

For any particular element in

$$\begin{vmatrix} a & d & g \\ b & e & h \\ c & f & i \end{vmatrix}$$

all other elements *not* in the same row or column as the element itself form a 2 x 2 determinant, called a **minor**. Thus, for the determinant in hand, we have the situation shown in the first two columns of Table 14.1. Moreover, if each minor is given a positive or negative sign according to the scheme shown in the third column of the table, then the resulting 'signed minors' are called **co-factors** and are shown in the last column of Table 14.1.

Now select any row or column of $|A|$, multiply each of the three elements in that row or column by their corresponding co-factors, and add the three products together: the result is $|A|$. If we select the first column

Table 14.1 Details of minors and co-factors in a third order determinant.

Element	Minor	Pattern of signs for co-factor	Co-factor
a	$\begin{vmatrix} e & h \\ f & i \end{vmatrix}$		$\begin{vmatrix} e & h \\ f & i \end{vmatrix}$
b	$\begin{vmatrix} d & g \\ f & i \end{vmatrix}$		$-\begin{vmatrix} d & g \\ f & i \end{vmatrix}$
c	$\begin{vmatrix} d & g \\ e & h \end{vmatrix}$		$\begin{vmatrix} d & g \\ e & h \end{vmatrix}$
d	$\begin{vmatrix} b & h \\ c & i \end{vmatrix}$		$-\begin{vmatrix} b & h \\ c & i \end{vmatrix}$
e	$\begin{vmatrix} a & g \\ c & i \end{vmatrix}$	$\begin{matrix} + & - & + \\ - & + & - \\ + & - & + \end{matrix}$	$\begin{vmatrix} a & g \\ c & i \end{vmatrix}$
f	$\begin{vmatrix} a & g \\ b & h \end{vmatrix}$		$-\begin{vmatrix} a & g \\ b & h \end{vmatrix}$
g	$\begin{vmatrix} b & e \\ c & f \end{vmatrix}$		$\begin{vmatrix} b & e \\ c & f \end{vmatrix}$
h	$\begin{vmatrix} a & d \\ c & f \end{vmatrix}$		$-\begin{vmatrix} a & d \\ c & f \end{vmatrix}$
i	$\begin{vmatrix} a & d \\ b & e \end{vmatrix}$		$\begin{vmatrix} a & d \\ b & e \end{vmatrix}$

of the determinant, then we have

$$\begin{vmatrix} a & d & g \\ b & e & h \\ c & f & i \end{vmatrix} = a\begin{vmatrix} e & h \\ f & i \end{vmatrix} + b\left\{-\begin{vmatrix} d & g \\ f & i \end{vmatrix}\right\} + c\begin{vmatrix} d & g \\ e & h \end{vmatrix}$$

$$= a(ei - fh) - b(di - fg) + c(dh - eg)$$

$$= aei - afh - bdi + bfg + cdh - ceg$$

We could equally well have selected another row or column; thus, employing the middle row:

$$|A| = b \left\{ -\begin{vmatrix} d & g \\ f & i \end{vmatrix} \right\} + e \begin{vmatrix} a & g \\ c & i \end{vmatrix} + h \left\{ -\begin{vmatrix} a & d \\ c & f \end{vmatrix} \right\}$$

$$= -b(di - fg) + e(ai - cg) - h(af - cd)$$

$$= -bdi + bfg + aei - ceg - afh + cdh$$

If this expression for the determinant is compared with the previous one, they will be found to be identical.

EXAMPLE 14.6
Evaluate

$$\begin{vmatrix} -1 & 2 & -3 \\ 2 & -1 & 4 \\ 3 & 4 & 1 \end{vmatrix}$$

Utilizing the first column, we have

$$= -1 \begin{vmatrix} -1 & 4 \\ 4 & 1 \end{vmatrix} + 2 \left\{ -\begin{vmatrix} 2 & -3 \\ 4 & 1 \end{vmatrix} \right\} + 3 \begin{vmatrix} 2 & -3 \\ -1 & 4 \end{vmatrix}$$

$$= -1(-1 - 16) - 2(2 + 12) + 3(8 - 3) = 17 - 28 + 15 = 4$$

Determinants of higher order matrices are evaluated in an analogous fashion. For each element there is a corresponding co-factor, and the determinant is calculated by multiplying all the elements in one row or column by their co-factors, and summing the resulting products. The signs for the co-factors are shown in the following pattern:

$$\begin{vmatrix} + & - & + & - \\ - & + & - & + \\ + & - & + & - \\ - & + & - & + \end{vmatrix}$$

and so

$$\begin{vmatrix} a & e & i & m \\ b & f & j & n \\ c & g & k & p \\ d & h & l & q \end{vmatrix}$$

$$= a \begin{vmatrix} f & j & n \\ g & k & p \\ h & l & q \end{vmatrix} - b \begin{vmatrix} e & i & m \\ g & k & p \\ h & l & q \end{vmatrix} + c \begin{vmatrix} e & i & m \\ f & j & n \\ h & l & q \end{vmatrix} - d \begin{vmatrix} e & i & m \\ f & j & n \\ g & k & p \end{vmatrix}$$

Evidently, the labour of calculating a determinant increases very rapidly with increase in order: in the above example of a fourth order determinant, it is first resolved into four third order determinants, and then each of these has to be evaluated as previously.

MATRIX INVERSION

The adjoint matrix

If A is a square matrix of order greater than 2, then the adjoint of A (adj. A) is the transpose of the matrix of the co-factors of A. The formation of an adjoint matrix is best demonstrated by means of an example.

EXAMPLE 14.7
Find the adjoint of the matrix

$$\begin{bmatrix} -1 & 2 & -3 \\ 2 & -1 & 4 \\ 3 & 4 & 1 \end{bmatrix}$$

The co-factor of -1 is $\begin{vmatrix} -1 & 4 \\ 4 & 1 \end{vmatrix} = -1 - 16 = -17$

The co-factor of 2 is $-\begin{vmatrix} 2 & -3 \\ 4 & 1 \end{vmatrix} = -(2 + 12) = -14$

The co-factor of 3 is $\begin{vmatrix} 2 & -3 \\ -1 & 4 \end{vmatrix} = 8 - 3 = 5$

The co-factor of 2 is $-\begin{vmatrix} 2 & 4 \\ 3 & 1 \end{vmatrix} = -(2 - 12) = 10$

The co-factor of -1 is $\begin{vmatrix} -1 & -3 \\ 3 & 1 \end{vmatrix} = -1 + 9 = 8$

The co-factor of 4 is $-\begin{vmatrix} -1 & -3 \\ 2 & 4 \end{vmatrix} = -(-4 + 6) = -2$

The co-factor of -3 is $\begin{vmatrix} 2 & -1 \\ 3 & 4 \end{vmatrix} = 8 + 3 = 11$

The co-factor of 4 is $-\begin{vmatrix} -1 & 2 \\ 3 & 4 \end{vmatrix} = -(-4 - 6) = 10$

The co-factor of 1 is $\begin{vmatrix} -1 & 2 \\ 2 & -1 \end{vmatrix} = 1 - 4 = -3$

Thus the matrix of co-factors is

$$\begin{bmatrix} -17 & 10 & 11 \\ -14 & 8 & 10 \\ 5 & -2 & -3 \end{bmatrix}$$

The transpose of this is the adjoint of the original matrix

$$\begin{bmatrix} -17 & -14 & 5 \\ 10 & 8 & -2 \\ 11 & 10 & -3 \end{bmatrix}$$

Only square matrices have an adjoint, and the definition given at the beginning of this section applies to third and higher order matrices. In the case of a second order matrix, the adjoint matrix is given by reversing the elements of the leading diagonal and changing the sign of the remaining two elements, as previously described (page 272).

The inverse matrix

We have already seen how to invert a second order matrix: the adjoint matrix is divided by the determinant. The same procedure is adopted for a matrix of any order.

EXAMPLE 14.8
Find the inverse matrix of

$$\begin{bmatrix} -1 & 2 & -3 \\ 2 & -1 & 4 \\ 3 & 4 & 1 \end{bmatrix}$$

The adjoint of this matrix has already been found (Example 14.7):

$$\begin{bmatrix} -17 & -14 & 5 \\ 10 & 8 & -2 \\ 11 & 10 & -3 \end{bmatrix}$$

so also has the determinant (Example 14.6), which is 4. Hence the inverse of the matrix is

$$\frac{1}{4} \begin{bmatrix} -17 & -14 & 5 \\ 10 & 8 & -2 \\ 11 & 10 & -3 \end{bmatrix} = \begin{bmatrix} -4\frac{1}{4} & -3\frac{1}{2} & 1\frac{1}{4} \\ 2\frac{1}{2} & 2 & -\frac{1}{2} \\ 2\frac{3}{4} & 2\frac{1}{2} & -\frac{3}{4} \end{bmatrix}$$

The singular matrix

Consider the matrix

$$\begin{bmatrix} 4 & 3 \\ -4 & -3 \end{bmatrix}$$

The adjoint of this matrix is

$$\begin{bmatrix} -3 & -3 \\ 4 & 4 \end{bmatrix}$$

but the determinant is $(4)(-3) - (-4)(3) = -12 - (-12) = 0$. Since an inverse matrix is formed by dividing every element in the adjoint matrix by the determinant, then if the determinant is zero the inverse matrix cannot be formed. A matrix whose determinant is zero is called a **singular matrix**, and the matrix has no inverse even though it is square.

EXAMPLE 14.9
  Show that

$$\begin{bmatrix} 2 & 3 & -2 \\ 1 & -2 & 3 \\ 4 & -1 & 4 \end{bmatrix}$$

is a singular matrix. To show that the matrix is singular, we need merely to calculate the determinant and see that it is zero. So

$$\begin{vmatrix} 2 & 3 & -2 \\ 1 & -2 & 3 \\ 4 & -1 & 4 \end{vmatrix} = 2 \begin{vmatrix} -2 & 3 \\ -1 & 4 \end{vmatrix} - 1 \begin{vmatrix} 3 & -2 \\ -1 & 4 \end{vmatrix} + 4 \begin{vmatrix} 3 & -2 \\ -2 & 3 \end{vmatrix}$$

$$= 2(-8 + 3) - 1(12 - 2) + 4(9 - 4)$$

$$= -10 - 10 + 20 = 0$$

SOME APPLICATIONS OF MATRIX ALGEBRA

By far the most common applications of matrix algebra in biology are in the fields of population dynamics, and in the statistical analysis of multivariate populations. The latter is based on the theory dealt with in Chapter 3, and so relates to problems in taxonomy, genecology, and ecology. However, it is impossible to deal with the application of matrix algebra to the statistical analysis of multivariate populations here, as much intermediate statistical theory needs to be known. Instead, more straight-forward applications will be presented, beginning with a very general one not confined to biology.

The solution of simultaneous linear equations

Suppose we had to solve the following pair of simultaneous linear equations (linear because both x and y are of the first degree and are in separate terms)

$$3x - 2y = \ 5$$
$$8x - 3y = 30$$

The method you know already is an elimination process. Label the two equations (1) and (2), respectively, then multiply (1) by 3 and multiply (2) by 2, giving

$$9x - 6y = 15$$
$$16x - 6y = 60$$

Subtraction gives $7x = 45$, and so $x = 6^3/_7$. Substitute this value of x into one of the two original equations (e.g. the first) $(3)(45/7) - 2y = 5$ which gives $y = 7^1/_7$.

Such equations can be solved by matrix methods, or even by using determinants alone.

Matrix solution

Using the above example, the pair of equations is

$$3x - 2y = \ 5$$
$$8x - 3y = 30$$

Now this pair of equations can be written in matrix form:

$$\begin{bmatrix} 3 & -2 \\ 8 & -3 \end{bmatrix} \begin{bmatrix} x \\ y \end{bmatrix} = \begin{bmatrix} 5 \\ 30 \end{bmatrix} \tag{14.20}$$

$(2 \times 2)(2 \times 1)$

Con X

as can be seen if the matrix multiplication on the left-hand side is carried out. Now, call the matrix of coefficients **A**, the column vector of the two unknowns **x**, and the column vector on the right-hand side **h**; then we have

$$\mathbf{A} \cdot \mathbf{x} = \mathbf{h} \tag{14.21}$$

where 14.21 is a more concise expression of 14.20. Next, pre-multiply both sides of 14.21 by the inverse of **A**:

$$\mathbf{A}^{-1} \cdot \mathbf{A} \cdot \mathbf{x} = \mathbf{A}^{-1} \cdot \mathbf{h}$$

i.e.

$$\mathbf{I} \cdot \mathbf{x} = \mathbf{A}^{-1} \cdot \mathbf{h}$$

$$\mathbf{I} = \begin{bmatrix} 1 & 0 \\ 0 & 1 \end{bmatrix}$$

since the product of a matrix and its inverse is the unit matrix. Further, as the product of any matrix and the unit matrix is the matrix itself (i.e. $\mathbf{I} \cdot \mathbf{x} = \mathbf{x}$), we have

$$\mathbf{x} = \mathbf{A}^{-1} \cdot \mathbf{h} \tag{14.22}$$

The left-hand side of 14.22 is the column vector of the two unknowns. The right-hand side consists of the product of a known column vector, **h**, and the inverse of the matrix of known coefficients, **A**, and so this side is known completely. Hence, the equation can be solved. Applying this to the above numerical example, we first require the inverse of

$$\begin{bmatrix} 3 & -2 \\ 8 & -3 \end{bmatrix}$$

The adjoint is

$$\begin{bmatrix} -3 & 2 \\ -8 & 3 \end{bmatrix}$$

and the determinant is $(-3)(3) - (-8)(2) = 7$; so the inverse is

$$\frac{1}{7} \begin{bmatrix} -3 & 2 \\ -8 & 3 \end{bmatrix}$$

Hence

$$\begin{bmatrix} x \\ y \end{bmatrix} = \frac{1}{7} \begin{bmatrix} -3 & 2 \\ -8 & 3 \end{bmatrix} \begin{bmatrix} 5 \\ 30 \end{bmatrix}$$

$$= \frac{1}{7} \begin{bmatrix} (-3)(5) + (2)(30) \\ (-8)(5) + (3)(30) \end{bmatrix} = \frac{1}{7} \begin{bmatrix} 45 \\ 50 \end{bmatrix} = \begin{bmatrix} 6\frac{3}{7} \\ 7\frac{1}{7} \end{bmatrix}$$

An advantage of the matrix method over the elimination procedure is that the same general formulation can be used for a group of simultaneous equations of any size, so long as the number of unknowns is equal to the number of equations in the set. However large the set, relationships 14.21 and 14.22 always apply.

EXAMPLE 14.10
    Solve, by the matrix method

$$x + 2y + z = 0$$

$$3x + 2y + z = 2$$

$$2x + 3y + 2z = 2$$

Put

$$\mathbf{A} = \begin{bmatrix} 1 & 2 & 1 \\ 3 & 2 & 1 \\ 2 & 3 & 2 \end{bmatrix} \quad \mathbf{x} = \begin{bmatrix} x \\ y \\ z \end{bmatrix} \quad \mathbf{h} = \begin{bmatrix} 0 \\ 2 \\ 2 \end{bmatrix}$$

The main problem is finding the inverse of **A**. When this has **been** achieved, pre-multiplication of h by $\mathbf{A}^{-1}$ gives the solution. Now

$$\text{adj} \cdot \mathbf{A} = \begin{bmatrix} 1 & -1 & 0 \\ -4 & 0 & 2 \\ 5 & 1 & -4 \end{bmatrix} \quad \text{and} \quad |A| = -2$$

Determinant

Hence

$$\mathbf{A}^{-1} = -\frac{1}{2} \begin{bmatrix} 1 & -1 & 0 \\ -4 & 0 & 2 \\ 5 & 1 & -4 \end{bmatrix}$$

and so

$$\begin{bmatrix} x \\ y \\ z \end{bmatrix} = -\frac{1}{2} \begin{bmatrix} 1 & -1 & 0 \\ -4 & 0 & 2 \\ 5 & 1 & -4 \end{bmatrix} \begin{bmatrix} 0 \\ 2 \\ 2 \end{bmatrix} = -\frac{1}{2} \begin{bmatrix} -2 \\ 4 \\ -6 \end{bmatrix} = \begin{bmatrix} 1 \\ -2 \\ 3 \end{bmatrix}$$

giving the solutions $x = 1, y = -2, z = 3$.

Determinant solution
    The method is best illustrated for a pair of equations, and we shall use

the previous example:

$$3x - 2y = 5$$

$$8x - 3y = 30$$

Define the following three determinants:

$$\Delta = \begin{vmatrix} 3 & -2 \\ 8 & -3 \end{vmatrix} \quad \Delta_1 = \begin{vmatrix} 5 & -2 \\ 30 & -3 \end{vmatrix} \quad \Delta_2 = \begin{vmatrix} 3 & 5 \\ 8 & 30 \end{vmatrix}$$

The first determinant is formed by the coefficients on the left-hand side of the equation. The second determinant has the coefficients of x in Δ replaced by the column of numbers from the right-hand side of the equations; whereas the third determinant has the coefficients of y in Δ replaced by this column.

A result, known as Cramer's Rule, states that

$$x = \Delta_1/\Delta \quad \text{and} \quad y = \Delta_2/\Delta$$

Now

$$\Delta = -9 + 16 = 7 \quad \Delta_1 = -15 + 60 = 45 \quad \Delta_2 = 90 - 40 = 50$$

Thus

$$x = \frac{45}{7} = 6\frac{3}{7} \quad \text{and} \quad y = \frac{50}{7} = 7\frac{1}{7} \quad \text{(as before)}$$

Cramer's Rule is applied in a similar fashion to a larger system of equations.

EXAMPLE 14.11

Solve the system of equations in Example 14.10, using Cramer's Rule. With three equations in three unknowns, we shall have third order determinants. Four of these will be required: the first being the determinant of the coefficients on the left-hand side,

$$\Delta = \begin{vmatrix} 1 & 2 & 1 \\ 3 & 2 & 1 \\ 2 & 3 & 2 \end{vmatrix} = -2$$

A second determinant is formed by substituting the values of the right-hand sides of the three equations for the first column of Δ,

$$\Delta_1 = \begin{vmatrix} 0 & 2 & 1 \\ 2 & 2 & 1 \\ 2 & 3 & 2 \end{vmatrix} = -2$$

The third determinant is formed by substituting in the second column of Δ,

$$\Delta_2 = \begin{vmatrix} 1 & 0 & 1 \\ 3 & 2 & 1 \\ 2 & 2 & 2 \end{vmatrix} = 4$$

Finally, a determinant is formed from Δ by substitution in the last column,

$$\Delta_3 = \begin{vmatrix} 1 & 2 & 0 \\ 3 & 2 & 2 \\ 2 & 3 & 2 \end{vmatrix} = -6$$

Then

$$x = \Delta_1/\Delta = -2/-2 = 1$$
$$y = \Delta_2/\Delta = 4/-2 = -2$$
$$z = \Delta_3/\Delta = -6/-2 = 3$$

(as before).

Population growth with age-dependent birth and death rates

We conclude with two biological applications of matrix algebra.

In considering the growth of a microbial population, we found that the logistic function could well describe the course of population growth with time. The relevant characteristic of the logistic function is that it allows for a slowing down of the growth rate with increasing density of the population, but the function does not permit other possible characteristics of population growth to be allowed for. In higher organisms, we have the situation that older individuals are more likely to die in a given interval of time than younger ones, and females of different ages may reproduce at different rates.

A model of population growth embodying these ideas can be formulated in terms of matrices. There are, however, two fundamental distinctions between this matrix model and the kind of model based on the logistic function. The latter regards time as a continuous variable, and so sets up a differential equation, whereas in the following matrix model time is thought of as a succession of discrete intervals (so that, of course, no differential equation can be formed). Thus the model is purely descriptive, and does not shed any light on the underlying dynamics of the population.

In what follows, the term 'individual' refers to any individual member of the population if the species is hermaphrodite, but only to a female member if the species is bisexual. In the latter case, however, the total numbers can be obtained if it is assumed that the ratio of the sexes, one to another, is constant, and that the death rates are the same for both sexes. At some time, t, a population can be represented by the column vector

$$\mathbf{n_t} = \begin{bmatrix} n_{0t} \\ n_{1t} \\ n_{2t} \\ \vdots \\ n_{mt} \end{bmatrix}$$

There are two subscripts to each element, and the second of these, t, is constant throughout and corresponds to the subscript of the symbol of the vector. The first subscript refers to the age of individuals in the population. Hence, at time t there are n_{0t} individuals between the ages of 0 and 1 units of time, n_{1t} individuals between the ages of 1 and 2, and so on. The sum of the elements gives the total number of individuals in the population. In general form, an element in the column vector $\mathbf{n_t}$ is n_{xt}, which denotes the number of individuals (at time t) between the ages of x and $x + 1$ units of time. There are $m + 1$ elements in the column vector, which assumes that no individual can survive longer than m units of time. We now enquire about the elements of the column vector $\mathbf{n_{t+1}}$, that is, the number of individuals of each age in the next time interval.

Denote by f_x the number of offspring (daughters) born to a female of age x that will survive into the next unit of time; the offspring will have age zero, and individuals already in existence at time t will move up to the next age group. The quantity f_x will, of course, be an average value. Next, denote by p_x the probability that an individual of age x will survive to the next time interval (probability is a number that can lie in the range $0 \leqslant p \leqslant 1$, where 0 denotes impossibility of an event, and 1 denotes certainty of an event). Write a matrix, $\mathbf{P}$, as follows:

$$\mathbf{P} = \begin{bmatrix} f_0 & f_1 & f_2 \cdots f_{m-1} & f_m \\ p_0 & 0 & 0 \ldots 0 & 0 \\ 0 & p_1 & 0 \ldots 0 & 0 \\ \vdots & \vdots & \vdots \quad \vdots & \vdots \\ 0 & 0 & 0 \ldots p_{m-1} & 0 \end{bmatrix}$$

Then the column vector, $\mathbf{n_{t+1}}$, describing the population at the next unit of time is given by

$$\mathbf{n_{t+1}} = \mathbf{P} \cdot \mathbf{n_t} \tag{14.23}$$

for we have

$$
\begin{bmatrix}
f_0 & f_1 & f_2 \cdots f_{m-1} & f_m \\
p_0 & 0 & 0 \ldots 0 & 0 \\
0 & p_1 & 0 \ldots 0 & 0 \\
\vdots & \vdots & \vdots \quad \vdots & \vdots \\
0 & 0 & 0 \ldots p_{m-1} & 0
\end{bmatrix}
\begin{bmatrix}
n_{0t} \\
n_{1t} \\
n_{2t} \\
\vdots \\
n_{mt}
\end{bmatrix}
$$

$$
= \begin{bmatrix}
f_0 n_{0t} + f_1 n_{1t} + \cdots + f_m n_{mt} \\
p_0 n_{0t} \\
p_1 n_{1t} \\
\vdots \\
p_{m-1} n_{(m-1)t}
\end{bmatrix}
$$

and so

$$
n_{t+1} = \begin{bmatrix}
n_{0(t+1)} \\
n_{1(t+1)} \\
n_{2(t+1)} \\
\vdots \\
n_{m(t+1)}
\end{bmatrix}
= \begin{bmatrix}
f_0 n_{0t} + f_1 n_{1t} + \cdots + f_m n_{mt} \\
p_0 n_{0t} \\
p_1 n_{1t} \\
\vdots \\
p_{m-1} n_{(m-1)t}
\end{bmatrix}
$$

The matrix P may be called a transition matrix.

For calculating population structure after further time units, we first have

$$
n_{t+2} = P \cdot n_{t+1} \tag{14.24}
$$

Substituting for n_{t+1} on the right-hand side of 14.24, using 14.23, gives

$$
n_{t+2} = P \cdot P \cdot n_t = P^2 \cdot n_t
$$

Hence, after s units of time,

$$
n_{t+s} = P^s \cdot n_t \tag{14.25}
$$

It is also possible to work backwards from the time, t, when the population was first observed. Matrix P is square, so should be invertable, to P^{-1}. Now pre-multiply both sides of 14.23 by P^{-1}:

$$
P^{-1} \cdot n_{t+1} = P^{-1} \cdot P \cdot n_t
$$

i.e.

$$n_t = P^{-1} \cdot n_{t+1}$$

So, commencing at time t:

$$n_{t-1} = P^{-1} \cdot n_t \qquad (14.26)$$

and

$$n_{t-2} = P^{-1} \cdot n_{t-1}$$
$$= P^{-1} \cdot P^{-1} \cdot n_t$$

or

$$= (P^{-1})^2 \cdot n_t$$

Hence, after s previous units of time

$$n_{t-s} = (P^{-1})^s \cdot n_t \qquad (14.27)$$

The values of the elements of the transition matrix are obtained by observation and experiment on the species concerned; or, they may be theoretical values, against which an actual population can be assessed by comparing the age structure of the population with column vectors produced by the theoretical model.

Species interaction in an ecosystem

Mathematically, this is a similar situation to the preceding one. Consider an ecosystem to have m co-existing species; at time t the number of each species present can be given by a column vector

$$n_t = \begin{bmatrix} n_{1t} \\ n_{2t} \\ n_{3t} \\ \vdots \\ n_{mt} \end{bmatrix}$$

Define a transition matrix, P, of order m so that the column vector of species numbers one unit of time later is given by

$$n_{t+1} = P \cdot n_t$$

To interpret the meaning of P, consider a specific, but highly artificial, example of an ecosystem with three species. Let

$$P = \begin{bmatrix} 1 + \dfrac{t}{20} & 0 & 0 \\ 0 & 1 & \dfrac{t}{20} \\ 0 & 0 & 1 - \dfrac{t}{10} \end{bmatrix}$$

Then

$$\mathbf{n_{t+1}} = \begin{bmatrix} 1 + \dfrac{t}{20} & 0 & 0 \\ 0 & 1 & \dfrac{t}{20} \\ 0 & 0 & 1 - \dfrac{t}{10} \end{bmatrix} \begin{bmatrix} n_{1t} \\ n_{2t} \\ n_{3t} \end{bmatrix} = \begin{bmatrix} n_{1t}\left(1 + \dfrac{t}{20}\right) \\ n_{2t} + \dfrac{n_{3t}t}{20} \\ n_{3t}\left(1 - \dfrac{t}{10}\right) \end{bmatrix}$$

It can be seen that species 1 increases in numbers independently of the other two. Species 3 decreases independently of the other two: whereas the rate of increase of species 2 depends on the current number of individuals of species 3. If P remains the same, eventually species 3 will become extinct and the number of individuals of species 2 will stabilize.

A biological interpretation of P can now be made. If P is a diagonal matrix, then each species changes independently of the others. It is the presence of off-diagonal elements that introduces species interaction. If P were a unit matrix, there would be no change in the species composition. In this example, species 2 changes in numbers only because species 3 does; species 2 is *partially* dependent upon species 3, since the former increases before the latter becomes extinct and then remains unchanged.

EXERCISES 14

1. If

$$P = \begin{bmatrix} 1 & 2 & -1 \\ -3 & -2 & 1 \\ 2 & 3 & 2 \end{bmatrix} \quad \text{and} \quad Q = \begin{bmatrix} 3 & 2 & -1 \\ 4 & 2 & -2 \\ -1 & 3 & 1 \end{bmatrix}$$

find (a) $P + Q$ (b) $Q + P$ (c) $P - Q$ (d) $Q - P$

2. If

$$A = \begin{bmatrix} 2 & 3 \\ 1 & 6 \end{bmatrix} \quad \text{and} \quad B = \begin{bmatrix} 1 & 2 \\ 3 & 4 \end{bmatrix}$$

find (a) $A \cdot B$ (b) $B \cdot A$

3. If

$$P = \begin{bmatrix} -1 & 2 \\ 2 & -4 \end{bmatrix} \text{ and } A \text{ and } B \text{ are as in Example 2}$$

find (a) $(A \cdot B) \cdot P$ (b) $A \cdot (B \cdot P)$

4. If

$$A = \begin{bmatrix} 3 & 2 & 1 \\ -2 & 3 & 6 \\ 1 & -2 & 4 \end{bmatrix} \quad \text{and} \quad b = \begin{bmatrix} 3 \\ 4 \\ 1 \end{bmatrix}$$

find, if possible, (a) $A \cdot b$ (b) $b \cdot A$

5.

$$\text{If } A \text{ is as in Example 4 and } I = \begin{bmatrix} 1 & 0 & 0 \\ 0 & 1 & 0 \\ 0 & 0 & 1 \end{bmatrix}$$

find (a) $A \cdot I$ (b) $I \cdot A$

6. If

$$C = \begin{bmatrix} 4 & 0 & 2 & 1 \\ -1 & 6 & 3 & 8 \end{bmatrix} \quad D = \begin{bmatrix} 2 & 3 \\ 4 & 1 \\ 6 & 8 \\ 9 & 2 \end{bmatrix}$$

$$E = \begin{bmatrix} 2 & 1 & 0 & 3 \\ 3 & 2 & 1 & 6 \\ -1 & 2 & 0 & 0 \\ 4 & 3 & 1 & 2 \end{bmatrix} \quad F = \begin{bmatrix} 2 & 1 & 2 \\ 3 & 0 & 6 \\ 2 & 1 & 8 \\ 4 & 1 & 6 \end{bmatrix}$$

find, if possible, (a) $C \cdot D$ (b) $C \cdot E$ (c) $C \cdot F$
                   (d) $D \cdot E$ (e) $E \cdot D$ (f) $E \cdot F$

7. Evaluate

(a)
$$\begin{vmatrix} 4 & -6 \\ 2 & 5 \end{vmatrix} \quad \text{and} \quad \text{(b)} \begin{vmatrix} 1 & 0 & 2 \\ 3 & 4 & 5 \\ 5 & 6 & 7 \end{vmatrix}$$

8. Let

$$A = \begin{bmatrix} 1 & 2 & 1 \\ 2 & 4 & 6 \\ 3 & 1 & 2 \end{bmatrix} \quad \text{and} \quad B = \begin{bmatrix} -1 & 2 & -3 \\ 2 & -1 & 4 \\ 3 & 4 & 1 \end{bmatrix}$$

Find (a) $A \cdot B$ (b) $B \cdot A$ (c) $|A|$
(d) $|B|$ (e) $|A \cdot B|$ (f) $|B \cdot A|$

What do these results show?

9. Find the adjoint matrices of A and B in Exercise 2 above. Hence, knowing the determinants, write down the matrices A^{-1} and B^{-1}.
10. Evaluate $|A^{-1}|$ and $|B^{-1}|$ of the previous example. What do these results show?

Answers to Exercises

Chapter 2

1. 1, 2, 6, 24, 120, 720, 5040, 40 320, 362 880, 3 628 800
2. (a) 90   (b) 220
3. (a) 8.9084   (b) −3.1282   (c) −9.9035
4. 0.2368
5. Put $\log_a b = x$ and $\log_b a = y$, then $b = a^x$ and $a = b^y$. Substituting the first of the two latter expressions into the second gives $a = a^{xy}$. Take logs of both sides:

$$\log a = xy \cdot \log a, \text{ thus } xy = 1, \text{ and the result is proved.}$$

6. (a) Put $\log_a a = x$ then $a = a^x$. Thus $x = 1$.
   (b) Since $\log_a a^x = x \cdot \log_a a$, and $\log_a a = 1$, the result is proved.

Chapter 3

1. (a) 7.616   (b) 4.583   (c) 7.348
2. This is best performed as a Class Exercise, and the results may be written in the form of a two-way table:

	V. beccabunga	*V. officinalis*	*V. montana*	*V. chamaedrys*	*V. serpyllifolia*	*V. hederifolia*	*V. persica*
V. officinalis	53.0						
V. montana	34.1	28.8					
V. chamaedrys	43.7	13.4	18.8				
V. serpyllifolia	77.5	27.1	55.7	38.9			
V. hederifolia	70.8	20.4	48.3	31.8	9.4		
V. persica	67.4	15.6	41.0	25.0	18.5	12.7	
V. filiformis	69.7	18.2	42.5	26.8	19.1	15.9	7.5

3. (a) $y = -2.1146 + 1.7321x$ (b) $y = 0.6667 + 0.6667x$

Chapter 4

1. (a) 1   (b) 13   (c) -35
2. (a) 34   (b) 16
3. (a) 0.1250   (b) 0.2857 mg CO_2 cm^{-2} hr^{-1}

Chapter 5

2. (a) $9(3x - 2)^2$ (b) $-(2ax + b)/(ax^2 + bx + c)^2$
   (c) $n(1 - 1/x^2)(x + 1/x)^{(n-1)}$
3. (a) $4x/(1 - x^2)^2$ (b) $(2 - x)/\{2\sqrt{(1 - x)^3}\}$
   (c) $x(2a^2 - 3x^2)/\sqrt{(a^2 - x^2)}$
   (d) $\{n(ad - bc)/(cx + d)^2\}\{(ax + b)/(cx + d)\}^{(n-1)}$
4. (a) $-1/x^2, 2/x^3$ (b) $ac^x \cdot \log_e c, ac^x(\log_e c)^2$

Chapter 6

1. For drawing the graphs, the following table provides some data in the range of n specified. For accurate graphs, more points will be required.

$0.2n$	$1 - n/1\,000\,000$	$0.2n(1 - n/1\,000\,000)$	$1 - (n/1\,000\,000)^2$	$0.2n\{1 - (n/1\,000\,000)^2\}$
100 000 20 000	0.9	18 000	0.99	19 800
200 000 40 000	0.8	32 000	0.96	38 400
300 000 60 000	0.7	42 000	0.91	54 600
400 000 80 000	0.6	48 000	0.84	67 200
500 000 100 000	0.5	50 000	0.75	75 000
600 000 120 000	0.4	48 000	0.64	76 800
700 000 140 000	0.3	42 000	0.51	71 400
800 000 160 000	0.2	32 000	0.36	57 600
900 000 180 000	0.1	18 000	0.19	34 200

Combination of equations 6.20 and 6.21 gives

$$\frac{dn}{dt} = 0.2n \left(1 - \frac{n}{1\,000\,000} \right)$$

and the co-ordinates of the maximum point are (500 000, 50 000).

Combination of equations 6.20 and 6.22 gives

$$\frac{dn}{dt} = 0.2n \left\{ 1 - \left(\frac{n}{1\,000\,000} \right)^2 \right\}$$

and the co-ordinates of the maximum point are (577 350, 76 980)

2. Maximum point: $(-1, 3)$ Minimum point: $(1, -1)$
Point of inflexion: $(0, 1)$
3. Discrimant is $(-3)^3 + 27(3)^2 = -27 + 243 = 216$, which is positive. Hence the equation has only one real root. By sketching the curve of the graph, the root is found to be near -2. Putting $X_0 = -2$, then $X_1 = -2.1$, $X_2 = -2.103836$, $X_3 = -2.103803$. Hence the real root of the equation is -2.1038 correct to 4 decimal places.

Chapter 7

1. (a) $2x^3 + c$ (b) $-5/x + c$ (c) $x^4/4 - a^3x + c$
(d) $x^3/3 - 2x - 1/x + c$ (e) $2\sqrt{x^3}/3 + 4\sqrt{x^5}/5 + 2\sqrt{x^7}/7 + c$
(f) $(2x - 1)^6/12 + c$ (g) $-3 \cdot \sqrt[3]{(1 - 4x)^2}/8 + c$
2. (a) $2|\cdot\log_e(x - 2) + \log_e(x + 3) + c$
(b) $\frac{1}{2}\cdot\log_e(x - 1) - \log_e(x - 2) + \frac{3}{2}\cdot\log_e(x - 3) + c$
3. (a) The substitution yields $\int dz/(z^2 - 1)$. This is then resolved into partial fractions, making use of the fact that $(z - 1)(z + 1) = z^2 - 1$. The final answer is

$$\tfrac{1}{2}[\log_e\{\sqrt{(1 + x^2)} - 1\} - \log_e\{\sqrt{(1 + x^2)} + 1\}] + c$$

(b) The substitution yields $\frac{1}{4}\cdot\int(z - 1)dz/z^3$, and the final answer is

$$\tfrac{1}{4}[1/\{2(2x + 1)^2\} - 1/(2x + 1)] + c$$

Chapter 8

1. (a) $16\frac{2}{3}$ (b) $14\frac{2}{3}$
2. (a) The answer will vary, depending on the accuracy of the graph and the counting technique, but should approximate to 0.5 square units.
(b) Working to 8 decimal places 0.50041761
(c) 0.50001247   (d) 0.5 exactly
3. (a) $8.\dot{3}$ (b) $2.9\dot{3}$
4. (a) Integral is meaningless   (b) 0.5

Chapter 9

1. From the information given, we have $a + 2d = -11$ and $a + 6d = 5$. Solving this pair of simultaneous equations yields $a = -19$ and $d = 4$; thus the series is $-19, -15, -11, -7, -3, 1, 5, 9, \ldots$ The twentieth term is 57, and $S_{20} = 380$.
2. (a) $x_1^2 + x_2^2 + x_3^2 + x_4^2 + x_5^2$
(b) $a(x^0 + x^1 + x^2 + x^3) = a(1 + x + x^2 + x^3)$
(c) $\sqrt{x_1}/(a + x_1) + \sqrt{x_2}/(a + x_2) + \sqrt{x_3}/(a + x_3) + \sqrt{x_4}/(a + x_4)$
(d) $x_1 . x_2 . x_3 . x_4$

3. (a) $\frac{1}{3} \sum\limits_{i=1}^{3} x_i$ (b) $\sum\limits_{i=2}^{-2} ix^{(i+3)}$ (c) $\frac{1}{3} \prod\limits_{i=1}^{3} x_i$ (d) $\sqrt[3]{\prod\limits_{i=1}^{3} x_i}$

4. From the information given $a/(1-r) = 36$ and $ar = 8$. These simultaneous equations give two solutions: either $a = 24$ and $r = 1/3$, or $a = 12$ and $r = 2/3$. So, two series fulfil the conditions:

$$24, 8, 8/3, 8/9, 8/27, \ldots$$
or $$12, 8, 16/3, 32/9, 64/27, \ldots$$

5. (a) 1/9   (b) 27/99   (c) 123/999
6. (a) $x^5 + 5x^4 + 10x^3 + 10x^2 + 5x + 1$
   (b) $x^5 - 5x^4 + 10x^3 - 10x^2 + 5x - 1$
   (c) $81x^4 - 216x^3y + 216x^2y^2 - 96xy^3 + 16y^4$
   (d) $x^8 + 12x^6y^2 + 54x^4y^4 + 108x^2y^6 + 81y^6$
7. (a) $1 + x + x^2 + x^3 + \ldots$
   (b) $2 + x^2/4 - x^4/64 + x^6/512 + \ldots$
   (c) $4 - x^3/3 - x^6/144 - x^9/2592 - \ldots$
8. (a) 1.00995   (b) 1.41421: expand $(1.9 + 0.1)^{1/2}$
9. (a) 0.6931   (b) -0.2231
   (c) $50° = 0.87266463$ rad, so $\sin 50° = 0.7761$
   (d) $20° = 0.34906585$ rad, so $\cos 20° = 0.9397$

Chapter 10

1. (a) 1.3499   (b) 0.8187
2. (a) Birch: 0.3041   Sycamore: 0.2243 day^{-1}
   (b) Birch: 0.0441   Sycamore: 0.3837 g
   (c) Birch: 10.5   Sycamore: 21.8 g
   (d) 27.1 days
3. 14.2 days
4. (a) 13.9   (b) 12.0 mm
5. (a) $be^{(a+bx)}$ (b) $2(x-1)e^{(1-2x+x^2)}$ (c) $-bke^{-kx}$
   (d) $(b + 2cx)/(a + bx + cx^2)$
6. (a) $\log_e(1+x) + c$ (b) $\frac{1}{3} \cdot \log_e(3x-7) + c$ (c) $t - be^{-kt}/k + c$
7. (a) 230   (b) 29   (c) 102   (d) 47
8. (a) $e^x(x-1) + c$ (b) $e^x(x^2 - 2x + 2) + c$
   (c) $\frac{1}{2} \cdot x^2(\log_e x - \frac{1}{2}) + c$ (d) $-\{(\log_e x)/x + 1/x\} + c$
9. (a) 1   (b) 1

Chapter 11

t	L_A/L_W cm² g⁻¹		L_W/W		$\bar{R}$ day⁻¹		$\bar{E}$ g cm⁻² d⁻¹	
	Unshaded	Shaded	Unshaded	Shaded	Unshaded	Shaded	Unshaded	Shaded
0	333.3	727.6	0.7500	0.6364	0.1703	0.1034	0.000681	0.000213
3	357.1	690.6	0.7000	0.7333	0.2189	0.0000	0.000776	0.000000
7	399.7	753.9	0.7708	0.7333	−0.0445	0.1834	−0.000200	0.000366
10	446.3	629.5	0.7381	0.7308	0.2067	0.1533	0.000729	0.000324
14	345.4	684.2	0.7292	0.7083	0.1465	0.0959	0.000612	0.000220
17	328.7	573.3	0.6980	0.6875	0.1758	−0.0215	0.000707	0.000037
21	389.0	684.4	0.6811	0.7167	−0.0839	0.1897	−0.000327	0.000392
24	370.7	684.6	0.6709	0.6981	0.1690	0.0678	0.000714	0.000149
28	331.9	606.0	0.6848	0.7194				

Chapter 12

1. (a) 1.0472   (b) 1.7453   (c) 0.2618 rad
2. (a) 5° 44'   (b) 154° 42'   (c) 80° 13'
3. (a) 0.9063   (b) 1.3456   (c) 2.7475   (d) 0.7660   (e) −0.3420
   (f) −1.1918   (g) −1.1034   (h) −0.0523   (i) 0.3249
   (j) −1.3902   (k) 0.5000   (l) −3.2709   (m) 0.3090   (n) 3.8637
   (o) 5.6713
4. (a) $2x(x \cdot \cos 2x + \sin 2x)$ (b) $(\tan x - x \cdot \sec^2 x)/\tan^2 x$
   (c) $3 \cdot \sin(2 - 3x)$
5. $dy/dx = 10e^{0.2 \cdot \sin x}(\cos x)$. Equating to zero yields the result $\cos x = 0$;
   so $x = \pi/2$ or $3\pi/2$ in the required range. Substituting these values
   successively into the original equation gives $y = 61.07$ or 40.94,
   respectively. $d^2y/dx^2 = 2e^{0.2 \cdot \sin x}(\cos^2 x - \sin x)$. When $x = \pi/2$,
   d^2y/dx^2 is negative; hence the point $(\pi/2, 61.07)$ is a maximum. When
   $x = 3\pi/2$, d^2y/dx^2 is positive; hence the point $(3\pi/2, 40.94)$ is a
   minimum.
6. (a) $-\frac{1}{4} \cdot \cos 4x + c$ (b) $-\frac{1}{2} \cdot e^{-x^2} + c$ (c) $\frac{1}{2} \cdot \sin^2 x + c$
7. (a) $x \cdot \sin x + \cos x + c$ (b) $\frac{1}{2} \cdot e^x(\sin x - \cos x) + c$

Chapter 13

1. (a) $y = a(1 - be^{-x/a})$ where $b = e^{-c/a}$, and c is the constant of
   integration.
   (b) The integral involving x has to be split into partial fractions, and so
   the differential equation emerges in the form $dy/(y - 1) = dx/x - dx/(x + 1)$. The final solution is $y = \{x(c + 1) + 1\}/(x + 1)$. The
   constant of integration is $\log_e c$.
   (c) The integral involving y has to be split into partial fractions,
   yielding $dy/(1 - y) + dy/(1 + y) = dx/x$. The final solution is
   $(1 + y)/(1 - y) = cx$, $\log_e c$ being the constant of integration.
   (d) Separating the variables gives $(\text{cosec } y)dy = -e^{-x}dx$. Integrating
   and re-arranging gives $\tan \frac{1}{2}y = bee^{-x}$; where $b = e^c$, c being the
   constant of integration. This is interpreted as follows: the tangent of
   the angle $\frac{1}{2}y$ (in radians) is bee^{-x}. This is written as $\frac{1}{2}y = \tan^{-1}(bee^{-x})$,
   ie $y = 2 \cdot \tan^{-1}(bee^{-x})$.
2. (a) Applying the integrating factor gives $y = e^{-2x}\int xe^{2x}dx + c$.
   Integrating by parts gives the final solution $y = \frac{1}{2}x - \frac{1}{4}$.
   (b) Applying the integrating factor gives

$$y = 2(\text{cosec } x) \int \cos 2x \cdot \sin x \cdot dx + c.$$

Integrating by parts twice gives

$$y = \frac{2}{3} \cdot \text{cosec } x\{(\cos 2x)(\cos x) + 2(\sin 2x)(\sin x)\}$$

(c) The equation first has to be divided throughout by $\cos x$, giving $dy/dx + y \cdot \tan x = \sec x$. The solution is $y = \sin x + c \cos x$

Chapter 14

1. (a) and (b) $\begin{bmatrix} 4 & 4 & -2 \\ 1 & 0 & -1 \\ 1 & 6 & 3 \end{bmatrix}$ (c) $\begin{bmatrix} -2 & 0 & 0 \\ -7 & -4 & 3 \\ 3 & 0 & 1 \end{bmatrix}$

(d) $\begin{bmatrix} 2 & 0 & 0 \\ 7 & 4 & -3 \\ -3 & 0 & -1 \end{bmatrix}$

2. (a) $\begin{bmatrix} 11 & 16 \\ 19 & 26 \end{bmatrix}$ (b) $\begin{bmatrix} 4 & 15 \\ 10 & 33 \end{bmatrix}$

3. (a) and (b) $\begin{bmatrix} 21 & -42 \\ 33 & -66 \end{bmatrix}$

4. (a) $\begin{bmatrix} 18 \\ 12 \\ -1 \end{bmatrix}$ (b) not conformable for multiplication

5. (a) and (b) $\begin{bmatrix} 3 & 2 & 1 \\ -2 & 3 & 6 \\ 1 & -2 & 4 \end{bmatrix}$

6. (a) $\begin{bmatrix} 29 & 30 \\ 112 & 43 \end{bmatrix}$ (b) $\begin{bmatrix} 10 & 11 & 1 & 14 \\ 45 & 41 & 14 & 49 \end{bmatrix}$ (c) $\begin{bmatrix} 16 & 7 & 30 \\ 54 & 10 & 106 \end{bmatrix}$

(d) not conformable for multiplication.

(e) $\begin{bmatrix} 35 & 13 \\ 74 & 31 \\ 6 & -1 \\ 44 & 27 \end{bmatrix}$ (f) $\begin{bmatrix} 19 & 5 & 28 \\ 38 & 10 & 62 \\ 4 & -1 & 10 \\ 27 & 7 & 46 \end{bmatrix}$

7. (a) 32   (b) −6

8. (a) $\begin{bmatrix} 6 & 4 & 6 \\ 24 & 24 & 16 \\ 5 & 13 & -3 \end{bmatrix}$ (b) $\begin{bmatrix} -6 & 3 & 5 \\ 12 & 4 & 4 \\ 14 & 23 & 29 \end{bmatrix}$

(c) 20   (d) 4

(e) The determinant of the product matrix $\mathbf{A} \cdot \mathbf{B}$ (in (a)) is 80

(f) The determinant of the product matrix $\mathbf{B} \cdot \mathbf{A}$ (in (b)) is 80. This shows that $|\mathbf{A} \cdot \mathbf{B}| = |\mathbf{B} \cdot \mathbf{A}| = |\mathbf{A}| \cdot |\mathbf{B}|$

9.

$$\text{Adj A} = \begin{bmatrix} 2 & -3 & 8 \\ 14 & -1 & -4 \\ -10 & 5 & 0 \end{bmatrix} \quad \text{Adj B} = \begin{bmatrix} -17 & -14 & 5 \\ 10 & 8 & -2 \\ 11 & 10 & -3 \end{bmatrix}$$

$$\text{A}^{-1} = \begin{bmatrix} \frac{1}{10} & -\frac{3}{20} & \frac{2}{5} \\ \frac{7}{10} & -\frac{1}{20} & -\frac{1}{5} \\ -\frac{1}{2} & \frac{1}{4} & 0 \end{bmatrix} \quad \text{B}^{-1} = \begin{bmatrix} -4\frac{1}{4} & -3\frac{1}{2} & 1\frac{1}{4} \\ 2\frac{1}{2} & 2 & -\frac{1}{2} \\ 2\frac{3}{4} & 2\frac{1}{2} & -\frac{3}{4} \end{bmatrix}$$

10. $|\text{A}^{-1}| = 0.05 \quad |\text{B}^{-1}| = 0.25$

This shows that the determinant of an inverse matrix is equal to the inverse of the determinant of the matrix itself, i.e. $|\text{A}^{-1}| = 1/|\text{A}| = |\text{A}|^{-1}$

LOGARITHMS

	0	1	2	3	4	5	6	7	8	9	1	2	3	4	5	6	7	8	9
10	0000	0043	0086	0128	0170	0212	0253	0294	0334	0374	4	8	12	17	21	25	29	33	37
11	0414	0453	0492	0531	0569	0607	0645	0682	0719	0755	4	8	11	15	19	23	26	30	34
12	0792	0828	0864	0899	0934	0969	1004	1038	1072	1106	3	7	10	14	17	21	24	28	31
13	1139	1173	1206	1239	1271	1303	1335	1367	1399	1430	3	6	10	13	16	19	23	26	29
14	1461	1492	1523	1553	1584	1614	1644	1673	1703	1732	3	6	9	12	15	18	21	24	27
15	1761	1790	1818	1847	1875	1903	1931	1959	1987	2014	3	6	8	11	14	17	20	22	25
16	2041	2068	2095	2122	2148	2175	2201	2227	2253	2279	3	5	8	11	13	16	18	21	24
17	2304	2330	2355	2380	2405	2430	2455	2480	2504	2529	2	5	7	10	12	15	17	20	22
18	2553	2577	2601	2625	2648	2672	2695	2718	2742	2765	2	5	7	9	12	14	16	19	21
19	2788	2810	2833	2856	2878	2900	2923	2945	2967	2989	2	4	7	9	11	13	16	18	20
20	3010	3032	3054	3075	3096	3118	3139	3160	3181	3201	2	4	6	8	11	13	15	17	19
21	3222	3243	3263	3284	3304	3324	3345	3365	3385	3404	2	4	6	8	10	12	14	16	18
22	3424	3444	3464	3483	3502	3522	3541	3560	3579	3598	2	4	6	8	10	12	14	15	17
23	3617	3636	3655	3674	3692	3711	3729	3747	3766	3784	2	4	6	7	9	11	13	15	17
24	3802	3820	3838	3856	3874	3892	3909	3927	3945	3962	2	4	5	7	9	11	12	14	16
25	3979	3997	4014	4031	4048	4065	4082	4099	4116	4133	2	3	5	7	9	10	12	14	15
26	4150	4166	4183	4200	4216	4232	4249	4265	4281	4298	2	3	5	7	8	10	11	13	15
27	4314	4330	4346	4362	4378	4393	4409	4425	4440	4456	2	3	5	6	8	9	11	13	14
28	4472	4487	4502	4518	4533	4548	4564	4579	4594	4609	2	3	5	6	8	9	11	12	14
29	4624	4639	4654	4669	4683	4698	4713	4728	4642	4757	1	3	4	6	7	9	10	12	13
30	4771	4786	4800	4814	4829	4843	4857	4871	4886	4900	1	3	4	6	7	9	10	11	13
31	4914	4928	4942	4955	4969	4983	4997	5011	5024	5038	1	3	4	6	7	8	10	11	12
32	5051	5065	5079	5092	5105	5119	5132	5145	5159	5172	1	3	4	5	7	8	9	11	12
33	5185	5198	5211	5224	5237	5250	5263	5276	5289	5302	1	3	4	5	6	8	9	10	12
34	5315	5328	5340	5353	5366	5378	5391	5403	5416	5428	1	3	4	5	6	8	9	10	11
35	5441	5453	5465	5478	5490	5502	5514	5527	5539	5551	1	2	4	5	6	7	9	10	11
36	5563	5575	5587	5599	5611	5623	5635	5647	5658	5670	1	2	4	5	6	7	8	10	11
37	5682	5694	5705	5717	5729	5740	5752	5763	5775	5786	1	2	3	5	6	7	8	9	10
38	5798	5809	5821	5832	5843	5855	5866	5877	5888	5899	1	2	3	5	6	7	8	9	10
39	5911	5922	5933	5944	5955	5966	5977	5988	5999	6010	1	2	3	4	5	7	8	9	10
40	6021	6031	6042	6053	6064	6075	6085	6096	6107	6117	1	2	3	4	5	6	8	9	10
41	6128	6138	6149	6160	6170	6180	6191	6201	6212	6222	1	2	3	4	5	6	7	8	9
42	6232	6243	6253	6263	6274	6284	6294	6304	6314	6325	1	2	3	4	5	6	7	8	9
43	6335	6345	6355	6365	6375	6385	6395	6405	6115	6425	1	2	3	4	5	6	7	8	9
44	6435	6444	6454	6464	6474	6484	6493	6503	6513	6522	1	2	3	4	5	6	7	8	9
45	6532	6542	6551	6561	6571	6580	6590	6599	6609	6618	1	2	3	4	5	6	7	8	9
46	6628	6637	6646	6656	6665	6675	6684	6693	6702	6712	1	2	3	4	5	6	7	7	8
47	6721	6730	6739	6749	6758	6767	6776	6785	6794	6803	1	2	3	4	5	5	6	7	8
48	6812	6821	6830	6839	6848	6857	6866	6875	6884	6893	1	2	3	4	4	5	6	7	8
49	6902	6911	6920	6928	6937	6946	6955	6964	6972	6981	1	2	3	4	4	5	6	7	8
50	6990	6998	7007	7016	7024	7033	7042	7050	7059	7067	1	2	3	3	4	5	6	7	8
51	7076	7084	7093	7101	7110	7118	7126	7135	7143	7152	1	2	3	3	4	5	6	7	8
52	7160	7168	7177	7185	7193	7202	7210	7218	7226	7235	1	2	2	3	4	5	6	7	7
53	7243	7251	7259	7267	7275	7284	7292	7300	7308	7316	1	2	2	3	4	5	6	6	7
54	7324	7332	7340	7348	7356	7364	7372	7380	7388	7396	1	2	2	3	4	5	6	6	7

LOGARITHMS

	0	1	2	3	4	5	6	7	8	9	1	2	3	4	5	6	7	8	9
55	7404	7412	7419	7427	7435	7443	7451	7459	7466	7474	1	2	2	3	4	5	5	6	7
56	7482	7490	7497	7505	7513	7520	7528	7536	7543	7551	1	2	2	3	4	5	5	6	7
57	7559	7566	7574	7582	7589	7597	7604	7612	7619	7627	1	2	2	3	4	5	5	6	7
58	7634	7642	7649	7657	7664	7672	7679	7686	7694	7701	1	1	2	3	4	4	5	6	7
59	7709	7716	7723	7731	7738	7745	7752	7760	7767	7774	1	1	2	3	4	4	5	6	7
60	7782	7789	7796	7803	7810	7818	7825	7832	7839	7846	1	1	2	3	4	4	5	6	6
61	7853	7860	7868	7875	7882	7889	7896	7903	7910	7917	1	1	2	3	4	4	5	6	6
62	7924	7931	7938	7945	7952	7959	7966	7973	7980	7987	1	1	2	3	3	4	5	6	6
63	7993	8000	8007	8014	8021	8028	8035	8041	8048	8055	1	1	2	3	3	4	5	5	6
64	8062	8069	8075	8082	8089	8096	8102	8109	8116	8122	1	1	2	3	3	4	5	5	6
65	8129	8136	8142	8149	8156	8162	8169	8176	8182	8189	1	1	2	3	3	4	5	5	6
66	8195	8202	8209	8215	8222	8228	8235	8241	8248	8254	1	1	2	3	3	4	5	5	6
67	8261	8267	8274	8280	8287	8293	8299	8306	8312	8319	1	1	2	3	3	4	5	5	6
68	8325	8331	8338	8344	8351	8357	8363	8370	8376	8382	1	1	2	3	3	4	4	5	6
69	8388	8395	8401	8407	8414	8420	8426	8432	8439	8445	1	1	2	2	3	4	4	5	6
70	8451	8457	8463	8470	8476	8482	8488	8494	8500	8506	1	1	2	2	3	4	4	5	6
71	8513	8519	8525	8531	8537	8543	8549	8555	8561	8567	1	1	2	2	3	4	4	5	5
72	8573	8579	8585	8591	8597	8603	8609	8615	8621	8627	1	1	2	2	3	4	4	5	5
73	8633	8639	8645	8651	8657	8663	8669	8675	8681	8686	1	1	2	2	3	4	4	5	5
74	8692	8698	8704	8710	8716	8722	8727	8733	8739	8745	1	1	2	2	3	4	4	5	5
75	8751	8756	8762	8768	8774	8779	8785	8791	8797	8802	1	1	2	2	3	3	4	5	5
76	8808	8814	8820	8825	8831	8837	8842	8848	8854	8859	1	1	2	2	3	3	4	5	5
77	8865	8871	8876	8882	8887	8893	8899	8904	8910	1815	1	1	2	2	3	3	4	4	5
78	8921	8927	8932	8938	8943	8949	8954	8960	8965	8971	1	1	2	2	3	3	4	4	5
79	8976	8982	8987	8993	8998	9004	9009	9015	9020	9025	1	1	2	2	3	3	4	4	5
80	9031	9036	9042	9047	9053	9058	9063	9069	9074	9079	1	1	2	2	3	3	4	4	5
81	9085	9090	9096	9101	9106	9112	9117	9122	9128	9133	1	1	2	2	3	3	4	4	5
82	9138	9143	9149	9154	9159	9165	9170	9175	9180	9186	1	1	2	2	3	3	4	4	5
83	9191	9196	9201	9206	9212	9217	9222	9227	9232	9238	1	1	2	2	3	3	4	4	5
84	9243	9248	9253	9258	9263	9269	9274	9279	9284	9289	1	1	2	2	3	3	4	4	5
85	9294	9299	9304	9309	9315	9320	9325	9330	9335	9340	1	1	2	2	3	3	4	4	5
86	9345	9350	9355	9360	9365	9370	9375	9380	9385	9390	1	1	2	2	3	3	4	4	5
87	9395	9400	9405	9410	9415	9420	9425	9430	9435	9440	0	1	1	2	2	3	3	4	4
88	9445	9450	9455	9460	9465	9469	9474	9479	9484	9489	0	1	1	2	2	3	3	4	4
89	9494	9499	9504	9509	9513	9518	9523	9528	9533	9538	0	1	1	2	2	3	3	4	4
90	9542	9547	9552	9557	9562	9566	9571	9576	9581	9586	0	1	1	2	2	3	3	4	4
91	9590	9595	9600	9605	9609	9614	9619	9624	9628	9633	0	1	1	2	2	3	3	4	4
92	9638	9643	9647	9652	9657	9661	9666	9671	9675	9680	0	1	1	2	2	3	3	4	4
93	9685	9689	9694	9699	9703	9708	9713	9717	9722	9727	0	1	1	2	2	3	3	4	4
94	9731	9736	9741	9745	9750	9754	9759	9763	9768	9773	0	1	1	2	2	3	3	4	4
95	9777	9782	9786	9791	9795	9800	9805	9809	9814	9818	0	1	1	2	2	3	3	4	4
96	9823	9827	9832	9836	9841	9845	9850	9854	9859	9863	0	1	1	2	2	3	3	4	4
97	9868	9872	9877	9881	9886	9890	9894	9899	9903	9908	0	1	1	2	2	3	3	4	4
98	9912	9917	9921	9926	9930	9934	9939	9943	9948	9952	0	1	1	2	2	3	3	4	4
99	9956	9961	9965	9969	9974	9978	9983	9987	9991	9996	0	1	1	2	2	3	3	3	4

ANTILOGARITHMS

	0	1	2	3	4	5	6	7	8	9	1	2	3	4	5	6	7	8	9
.00	1000	1002	1005	1007	1009	1012	1014	1016	1019	1021	0	0	1	1	1	1	2	2	2
.01	1023	1026	1028	1030	1033	1035	1038	1040	1042	1045	0	0	1	1	1	1	2	2	2
.02	1047	1050	1052	1054	1057	1059	1062	1064	1067	1069	0	0	1	1	1	1	2	2	2
.03	1072	1074	1076	1079	1081	1084	1086	1089	1089	1091	0	0	1	1	1	1	2	2	2
.04	1096	1099	1102	1104	1107	1109	1112	1114	1117	1119	0	1	1	1	1	2	2	2	2
.05	1122	1125	1127	1230	1132	1135	1138	1140	1143	1146	0	1	1	1	1	2	2	2	2
.06	1148	1151	1153	1156	1159	1161	1164	1167	1169	1172	0	1	1	1	1	2	2	2	2
.07	1175	1178	1180	1183	1186	1189	1191	1194	1197	1199	0	1	1	1	1	2	2	2	2
.08	1202	1205	1208	1211	1213	1216	1219	1222	1225	1227	0	1	1	1	1	2	2	2	3
.09	1230	1233	1236	1239	1242	1245	1247	1250	1253	1256	0	1	1	1	1	2	2	2	3
.10	1259	1262	1265	1268	1271	1274	1276	1279	1282	1285	0	1	1	1	1	2	2	2	3
.11	1288	1291	1294	1297	1300	1303	1306	1309	1312	1315	0	1	1	1	2	2	2	2	3
.12	1318	1321	1324	1327	1330	1334	1337	1340	1343	1346	0	1	1	2	2	2	2	2	3
.13	1349	1352	1355	1358	1361	1365	1368	1371	1374	1377	0	1	1	1	2	2	2	3	3
.14	1380	1384	1387	1390	1393	1396	1400	1403	1406	1409	0	1	1	1	2	2	2	3	3
.15	1413	1416	1419	1422	1426	1429	1432	1435	1439	1442	0	1	1	1	2	2	2	3	3
.16	1445	1449	1452	1455	1459	1462	1466	1469	1472	1476	0	1	1	1	2	2	2	3	3
.17	1479	1483	1486	1489	1493	1496	1500	1503	1507	1510	0	1	1	1	2	2	2	3	3
.18	1514	1517	1521	1524	1528	1531	1535	1538	1542	1545	0	1	1	1	2	2	2	3	3
.19	1549	1552	1556	1560	1563	1567	1570	1574	1578	1581	0	1	1	1	2	2	3	3	3
.20	1585	1589	1592	1596	1600	1603	1607	1611	1614	1618	0	1	1	1	2	2	3	3	3
.21	1622	1626	1629	1633	1637	1641	1644	1648	1652	1656	0	1	1	2	2	2	3	3	3
.22	1660	1663	1667	1671	1675	1679	1683	1687	1690	1694	0	1	1	2	2	2	3	3	3
.23	1698	1702	1706	1710	1714	1718	1722	1726	1730	1734	0	1	1	2	2	2	3	3	4
.24	1738	1742	1746	1750	1754	1758	1762	1766	1770	1774	0	1	1	2	2	2	3	3	3
.25	1778	1782	1786	1791	1795	1799	1803	1807	1811	1816	0	1	1	2	2	2	3	3	4
.26	1820	1824	1828	1832	1837	1841	1845	1849	1854	1858	0	1	1	2	2	3	3	3	4
.27	1862	1866	1871	1875	1879	1884	1888	1892	1897	1901	0	1	1	2	2	3	3	3	4
.28	1905	1910	1914	1919	1923	1928	1932	1936	1941	1945	0	1	1	2	2	3	3	4	4
.29	1950	1954	1959	1963	1968	1972	1977	1982	1986	1991	0	1	1	2	2	3	3	4	4
.30	1995	2000	2004	2009	2014	2018	1023	2028	2032	2037	0	1	1	2	2	3	3	4	4
.31	2042	2046	2051	2056	2061	2065	2070	2075	2080	2084	0	1	1	2	2	3	3	4	4
.32	2089	2094	2099	2104	2109	2113	2118	2123	2128	2133	0	1	1	2	2	3	3	4	4
.33	2138	2143	2148	2153	2158	2163	2168	2173	2178	2183	0	1	1	2	2	3	3	4	4
.34	2188	2193	2198	2203	2208	2213	2218	2223	2228	2234	1	1	2	2	3	3	4	4	5
.35	2239	2244	2249	2254	2259	2265	2270	2275	2280	2286	1	1	2	2	3	3	4	4	5
.36	2291	2296	2301	2307	2312	2317	2323	2328	2333	2339	1	1	2	2	3	3	4	4	5
.37	2344	2350	2355	2360	2366	2371	2377	2382	2388	2393	1	1	2	2	3	3	4	4	5
.38	2399	2404	2410	2415	2421	2417	2432	2438	2443	2449	1	1	2	2	3	3	4	4	5
.39	2455	2460	2466	2472	2477	2483	2489	2495	2500	2506	1	1	2	2	3	3	4	5	5
.40	2512	2518	2523	2529	2535	2541	2547	2553	2559	2564	1	1	2	2	3	4	4	5	5
.41	2570	2576	2582	2588	2594	2600	2606	2612	2618	2624	1	1	2	2	3	4	4	5	5
.42	2630	2636	2642	2649	2655	2661	2667	2673	2679	2685	1	1	2	2	3	4	4	5	6
.43	2692	2698	2704	2710	2716	2723	2729	2735	2742	2748	1	1	2	3	3	4	4	5	6
.44	2754	2761	2767	2773	2780	2786	2793	2799	2805	2812	1	1	2	3	3	4	4	5	6
.45	2818	2825	2831	2838	2844	2851	2858	2864	2871	2877	1	1	2	3	3	4	5	5	6
.46	2884	2891	2897	2904	2911	2917	1924	2931	2938	2944	1	1	2	3	3	4	5	5	6
.47	2951	2958	2965	2972	2979	2985	2992	2999	3006	3013	1	1	2	3	3	4	5	5	6
.48	3020	3027	3034	3041	3048	3055	3062	3069	3076	3083	1	1	2	3	4	4	5	6	6
.49	3090	3097	3105	3112	3119	3126	3133	3141	3148	3155	1	1	2	3	4	4	5	6	6

ANTILOGARITHMS

	0	1	2	3	4	5	6	7	8	9	1	2	3	4	5	6	7	8	9
.50	3162	3170	3177	3184	3192	3199	3206	3214	3221	3228	1	1	2	3	4	4	5	6	7
.51	3236	3243	3251	3258	3266	3273	3281	3289	3296	3304	1	2	2	3	4	5	5	6	7
.52	3311	3319	3327	3334	3342	3350	3357	3365	3373	3381	1	2	2	3	4	5	5	6	7
.53	3388	3396	3404	3412	3420	3428	3436	3443	3451	3459	1	2	2	3	4	5	6	6	7
.54	3467	3475	3483	3491	3499	3508	3516	3524	3532	3549	1	2	2	3	4	5	6	6	7
.55	3548	3556	3565	3573	3581	3589	3597	3606	3614	1622	1	2	2	3	4	5	6	7	7
.56	3631	3639	3648	3656	3664	3673	3681	3690	3698	3707	1	2	3	3	4	5	6	7	8
.57	3715	3724	3733	3741	3750	3758	3767	3776	3784	3793	1	2	3	3	4	5	6	7	8
.58	3802	3811	3819	3828	3937	3846	3855	3864	3873	3882	1	2	3	4	4	5	6	7	8
.59	3890	3899	3908	3917	3926	3936	3945	3954	3963	3972	1	2	3	4	5	5	6	7	8
.60	3981	3990	3999	4009	4018	4027	4036	4046	4055	4064	1	2	3	4	5	6	6	7	8
.61	4074	4083	4093	4102	4111	4121	5130	4140	4150	4159	1	2	3	4	5	6	7	8	9
.62	4169	4178	4188	4198	4207	4217	4227	4236	4246	4256	1	2	3	4	5	6	7	8	9
.63	4266	4276	4285	4295	4305	4315	4325	4335	4345	4355	1	2	3	4	5	6	7	8	9
.64	4365	4375	4385	4395	4406	4416	4426	4436	4446	4457	1	2	3	4	5	6	7	8	9
.65	4467	4477	4487	4498	4508	4519	4529	4539	4550	4560	1	2	3	4	5	6	7	8	9
.66	4571	4581	4592	4603	4613	4624	4634	4645	4656	4667	1	2	3	4	5	6	7	9	10
.67	4677	4688	4699	4710	4721	4732	4742	4753	4764	4775	1	2	3	4	5	7	8	9	10
.68	4786	4797	4808	4819	4831	4842	4853	4864	4875	4887	1	2	3	4	6	7	8	9	10
.69	4898	4909	4920	4932	4943	4955	4966	4977	4989	5000	1	2	3	5	6	7	8	9	10
.70	5012	5023	5035	5047	5058	5070	5082	5093	5105	5117	1	2	4	5	6	7	8	9	11
.71	5129	5140	5152	5164	5176	5188	5200	5212	5224	5236	1	2	4	5	6	7	8	10	11
.72	5248	5260	5272	5284	5297	5309	5321	5333	5346	5358	1	2	4	5	6	7	9	10	11
.73	5370	5383	5395	5408	5420	5433	5445	5458	5470	5483	1	3	4	5	6	8	9	10	11
.74	5495	5508	5521	5534	5546	5559	5572	5585	5598	5610	1	3	4	5	6	8	9	10	12
.75	5623	5636	5649	5662	5675	5689	5702	5715	5728	5741	1	3	4	5	7	8	9	10	12
.76	5754	5768	5781	5794	5808	5821	5834	5848	5861	5875	1	3	4	5	7	8	9	11	12
.77	5888	5902	5916	5929	5943	5957	5970	5984	5998	6012	1	3	4	5	7	8	10	11	12
.78	6026	6039	6053	6067	6081	6095	6109	6124	6138	6152	1	3	4	6	7	8	10	11	13
.79	6166	6180	6194	6209	6223	6237	6252	6266	6281	6295	1	3	4	6	7	9	10	11	13
.80	6310	6324	6339	6353	6368	6383	6397	6412	6427	6442	1	3	4	6	7	9	10	12	13
.81	6457	6471	6486	6501	6516	6531	6546	6561	6577	6592	2	3	5	6	8	9	11	12	14
.82	6607	6622	6637	6653	6668	6683	6699	6714	6730	6745	2	3	5	6	8	9	11	12	14
.83	6761	6776	6792	6808	6823	6839	6855	6871	6887	6902	2	3	5	6	8	9	11	13	14
.84	6918	6934	6950	6966	6982	6998	7015	7031	7047	7063	2	3	5	6	8	10	11	13	15
.85	7079	7096	7112	7129	7145	7161	7178	7194	7211	7228	2	3	5	6	8	10	12	13	15
.86	7244	7261	7278	7295	7311	7328	7345	7362	7379	7396	2	3	5	7	8	10	12	13	15
.87	7413	7430	7447	7464	7482	7499	7516	7534	7551	7568	2	3	5	7	9	10	12	14	16
.88	7586	7603	7621	7638	7656	7674	7691	7709	7727	7745	2	4	5	7	9	11	12	14	16
.89	7762	7780	7798	7816	7834	7852	7870	7889	7907	7925	2	4	5	7	9	11	13	14	16
.90	7943	7962	7980	7998	8017	8035	8054	8072	8091	8110	2	4	6	7	9	11	13	15	17
.91	8128	8147	8166	8185	8204	8222	8241	8260	8279	8299	2	4	6	8	9	11	13	15	17
.92	8318	8337	8356	8375	8395	8414	8433	8453	8472	8492	2	4	6	8	10	12	14	15	17
.93	8511	8531	8551	8570	8590	8610	8630	8650	8670	8690	2	4	6	8	10	12	14	16	18
.94	8710	8730	8750	8770	8790	8810	8831	8851	8872	8892	2	4	6	8	10	12	14	16	18
.95	8913	8933	8954	8974	8995	9016	9036	9057	9078	9099	2	4	6	8	10	12	15	17	19
.96	9120	9141	9162	9183	9204	9226	9247	9268	9290	9311	2	4	6	8	11	13	15	17	19
.97	9333	9354	9376	9397	9419	9441	9462	9484	9506	9528	2	4	7	9	11	13	15	17	20
.98	9550	9572	9594	9616	9638	9661	9683	9705	9727	9750	2	4	7	9	11	13	16	18	20
.99	9772	9795	9817	9840	9863	9886	9908	9931	9954	9977	2	5	7	9	11	14	16	18	20

NATURAL LOGARITHMS

	0	1	2	3	4	5	6	7	8	9	1	2	3	4	5	6	7	8	9
1.0	0.0000	0099	0198	0296	0392	0488	0583	0677	0770	0862	10	19	29	38	48	57	67	76	86
1.1	.0953	1044	1133	1222	1310	1398	1484	1570	1655	1740	9	17	26	35	44	52	61	70	78
1.2	.1823	1906	1989	2070	2151	2231	2311	2390	2469	2546	8	16	24	32	40	48	56	64	72
1.3	.2624	2700	2776	2852	2927	3001	3075	3148	3221	3293	7	15	22	30	37	44	52	59	67
1.4	.3365	3436	3507	3577	3646	3716	3784	3853	3920	3988	7	14	21	28	35	41	48	55	62
1.5	.4055	4121	4187	4253	4318	4383	4447	4511	4574	4637	6	13	19	26	32	39	45	52	58
1.6	.4700	4762	4824	4886	4947	5008	5068	5128	5188	5247	6	12	18	24	30	36	42	48	55
1.7	.5306	5365	5423	5481	5539	5596	5653	5710	5766	5822	6	11	17	24	29	34	40	46	51
1.8	.5878	5933	5988	6043	6098	6152	6206	6259	6313	6366	5	11	16	22	27	32	38	43	49
1.9	.6419	6471	6523	6575	6627	6678	6729	6780	6831	6881	5	10	15	20	26	31	36	41	46
2.0	.6931	6981	7031	7080	7129	7178	7227	7275	7324	7372	5	10	15	20	24	29	34	39	44
2.1	.7419	7467	7514	7561	7608	7655	7701	7747	7793	7839	5	9	14	19	23	28	33	37	42
2.2	.7885	7930	7975	8020	8065	8109	8154	8198	8242	8286	4	9	13	18	22	27	31	36	40
2.3	.8329	8372	8416	8459	8502	8544	8587	8629	8671	8713	4	9	13	17	21	26	30	34	38
2.4	.8755	8796	8838	8879	8920	8961	9002	9042	9083	9123	4	8	12	16	20	24	29	33	37
2.5	.9163	9203	9243	9282	9322	9361	9400	9439	9478	9517	4	8	12	16	20	24	27	31	35
2.6	.9555	9594	9632	9670	9708	9746	9783	9821	9858	9895	4	8	11	15	19	23	26	30	34
2.7	.9933	9969	1.0006	0043	0080	0116	0152	0188	0225	0260	4	7	11	15	18	22	25	28	33
2.8	1.0296	0332	0367	0403	0438	0473	0508	0543	0578	0613	4	7	11	14	18	21	25	28	32
2.9	1.0647	0682	0716	0750	0784	0818	0852	0886	0919	0953	3	7	10	14	17	20	24	27	31
3.0	1.0986	1019	1053	1086	1119	1151	1184	1217	1249	1282	3	7	10	13	16	20	23	26	30
3.1	1.1314	1346	1378	1410	1442	1474	1506	1537	1569	1600	3	6	10	13	16	19	22	25	29
3.2	1.1632	1663	1694	1725	1756	1787	1817	1848	1878	1909	3	6	9	12	15	18	22	25	28
3.3	1.1939	1969	1.2000	2030	2060	2090	2119	2149	2179	2208	3	6	9	12	15	18	21	24	27
3.4	1.2238	2267	2296	2326	2355	2384	2413	2442	2470	2499	3	6	9	12	15	17	20	23	26
3.5	1.2528	2556	2585	2613	2641	2669	2698	2726	2754	2782	3	6	8	11	14	17	20	23	25
3.6	1.2809	2837	2865	2892	2920	2947	2975	3002	3029	3056	3	5	8	11	14	16	19	22	25
3.7	1.3083	3110	3137	3164	3191	3218	3244	3271	3297	3324	3	5	8	11	13	16	19	21	24
3.8	1.3350	3376	3403	3429	3455	3481	3507	3533	3558	3584	3	5	8	10	13	16	18	21	23
3.9	1.3610	3635	3661	3686	3712	3737	3762	3788	3813	3838	3	5	8	10	13	15	18	20	23
4.0	1.3863	3888	3913	3938	3962	3987	4012	4036	4061	4085	2	5	7	10	12	15	17	20	22
4.1	1.4110	4134	4159	4183	4207	4231	4255	4279	4303	4327	2	5	7	10	12	14	17	19	22
4.2	1.4351	4375	4398	4422	4446	4469	4493	4516	4540	4563	2	5	7	9	12	14	16	19	21
4.3	1.4586	4609	4633	4656	4679	4702	4725	4748	4770	4793	2	5	7	9	12	14	16	18	21
4.4	1.4816	4839	4861	4884	4907	4929	4951	4974	4996	5019	2	5	7	9	11	14	16	18	20
4.5	1.5041	5063	5085	5107	5129	5151	5173	5195	5217	5239	2	4	7	9	11	13	15	18	20
4.6	1.5261	5282	5304	5326	5347	5369	5390	5412	5433	5454	2	4	6	9	11	13	15	17	19
4.7	1.5476	5497	5518	5539	5560	5581	5602	5623	5644	5665	2	4	6	8	11	13	15	17	19
4.8	1.5686	5707	5728	5748	5769	5790	5810	5831	5851	5872	2	4	6	8	10	12	14	16	19
4.9	1.5892	5913	5933	5953	5974	5994	6014	6034	6054	6075	2	4	6	8	10	12	14	16	18
5.0	1.6094	6114	6134	6154	6174	6194	6214	6233	6253	6273	2	4	6	8	10	12	14	16	18
5.1	1.6292	6312	6332	6351	6371	6390	6409	6429	6448	6467	2	4	6	8	10	12	14	16	18
5.2	1.6487	6506	6525	6544	6563	6582	6601	6620	6639	6658	2	4	6	8	11	13	15	17	19
5.3	1.6677	6696	6715	6734	6752	6771	6790	6808	6827	6845	2	4	6	7	9	11	13	15	17
5.4	1.6864	6882	6901	6919	6938	6956	6974	6993	7011	7029	2	4	5	7	9	11	13	15	17

To find the natural logarithm of a number greater than 10, factorise and add the natural logarithm of the appropriate power of 10 (found from the table below) to the appropriate value tabulated above.

x	1	2	3	4	5	6	7	8
$\log_e 10^x$	2.3026	4.6052	6.9078	9.2103	11.5129	13.8155	16.1181	18.4207

NATURAL LOGARITHMS

	0	1	2	3	4	5	6	7	8	9	1	2	3	4	5	6	7	8	9
5.5	1.7047	7066	7084	7102	7120	7138	7156	7174	7192	7210	2	4	5	7	9	11	13	14	16
5.6	1.7228	7246	7263	7281	7299	7317	7334	7352	7370	7387	2	4	5	7	9	11	12	14	16
5.7	1.7405	7422	7440	7457	7475	7492	7509	7527	7544	7561	2	3	5	7	9	10	12	14	16
5.8	1.7579	7596	7613	7630	7647	7664	7681	7699	7716	7733	2	3	5	7	9	10	12	14	15
5.9	1.7750	7766	7783	7800	7817	7834	7851	7867	7884	7901	2	3	5	7	8	10	12	13	15
6.0	1.7918	7934	7951	7967	7984	8001	8017	8034	8050	8066	2	3	5	7	8	10	12	13	15
6.1	1.8083	8099	8116	8132	8148	8165	8181	8197	8213	8229	2	3	5	6	8	10	11	13	15
6.2	1.8245	8262	8278	8294	8310	8326	8342	8358	8374	8390	2	3	5	6	8	10	11	13	14
6.3	1.8405	8421	8437	8453	8469	8485	8500	8516	8532	8547	2	3	5	6	8	9	11	13	14
6.4	1.8563	8579	8594	8610	8625	8641	8656	8672	8687	8703	2	3	5	6	8	9	11	12	14
6.5	1.8718	8733	8749	8764	8779	8795	8810	8825	8840	8856	2	3	5	6	8	9	11	12	14
6.6	1.8871	8886	8901	8916	8931	8946	8961	8976	8991	9006	2	3	5	6	8	9	11	12	14
6.7	1.9021	9036	9051	9066	9081	9095	9110	9125	9140	9155	1	3	4	6	7	9	10	12	13
6.8	1.9196	9184	9199	9213	9228	9242	9257	9272	9286	9301	1	3	4	6	7	9	10	12	13
6.9	1.9315	9330	9344	9359	9373	9387	9402	9416	9430	9445	1	3	4	6	7	9	10	12	13
7.0	1.9459	9473	9488	9502	9516	9530	9544	9559	9573	9587	1	3	4	6	7	9	10	11	13
7.1	1.9601	9615	9629	9643	9657	9671	9685	9699	9713	9727	1	3	4	6	7	8	10	11	13
7.2	1.9741	9755	9769	9782	9796	9810	9824	9838	9851	9865	1	3	4	6	7	8	10	11	12
7.3	1.9879	9892	9906	9920	9933	9947	9961	9974	9988	2.0001	1	3	4	5	7	8	10	11	12
7.4	2.0015	0028	0042	0055	0069	0082	0096	0109	0122	0136	1	3	4	5	7	8	9	11	12
7.5	2.0149	0162	0176	0189	0202	0215	0229	0242	0255	0268	1	3	4	5	7	8	9	11	12
7.6	2.0281	0295	0308	0321	0334	0347	0360	0373	0386	0399	1	3	4	5	7	8	9	10	12
7.7	2.0412	0425	0438	0451	0464	0477	0490	0503	0516	0528	1	3	4	5	6	8	9	10	12
7.8	2.0541	0554	0567	0580	0592	0605	0618	0631	0643	0656	1	3	4	5	6	8	9	10	11
7.9	2.0669	0681	0694	0707	0719	0732	0744	0757	0769	0782	1	3	4	5	6	8	9	10	11
8.0	2.0794	0807	0819	0832	0844	0857	0869	0882	0894	0906	1	3	4	5	6	7	9	10	11
8.1	2.0919	0931	0934	0956	0968	0980	0992	1005	1017	1029	1	2	4	5	6	7	9	10	11
8.2	2.1041	1054	1066	1078	1090	1102	1114	1126	1138	1150	1	2	4	5	6	7	9	10	11
8.3	2.1163	1175	1187	1199	1211	1223	1235	1247	1258	1270	1	2	4	5	6	7	8	10	11
8.4	2.1282	1294	1306	1318	1330	1342	1353	1365	1377	1389	1	2	4	5	6	7	8	9	11
8.5	2.1401	1412	1424	1436	1448	1459	1471	1483	1494	1506	1	2	4	5	6	7	8	9	11
8.6	2.1518	1529	1541	1552	1564	1576	1587	1599	1610	1622	1	2	3	5	6	7	8	9	10
8.7	2.1633	1645	1656	1668	1679	1691	1702	1713	1725	1736	1	2	3	5	6	7	8	9	10
8.8	2.1748	1759	1770	1782	1793	1804	1815	1827	1838	1849	1	2	3	5	6	7	8	9	10
8.9	2.1861	1872	1883	1894	1905	1917	1928	1939	1950	1961	1	2	3	4	6	7	8	9	10
9.0	2.1972	1983	1994	2006	2017	2028	2039	2050	2061	2072	1	2	3	4	6	7	8	9	10
9.1	2.2083	2094	2105	2116	2127	2138	2148	2159	2170	2181	1	2	3	4	5	7	8	9	10
9.2	2.2192	2203	2214	2225	2235	2246	2257	2268	2279	2289	1	2	3	4	5	6	8	9	10
9.3	2.2300	2311	2322	2332	2343	2354	2364	2375	2386	2396	1	2	3	4	5	6	7	9	10
9.4	2.2407	2418	2428	2439	2450	2460	2471	2481	2492	2502	1	2	3	4	5	6	7	8	10
9.5	2.2513	2523	2534	2544	2555	2565	2576	2586	2597	2607	1	2	3	4	5	6	7	8	9
9.6	2.2618	2628	2638	2649	2659	2670	2680	2690	2701	2711	1	2	3	4	5	6	7	8	9
9.7	2.2721	2732	2742	2752	2762	2773	2783	2793	2803	2814	1	2	3	4	5	6	7	8	9
9.8	2.2824	2834	2844	2854	2865	2875	2885	2895	2905	2915	1	2	3	4	5	6	7	8	9
9.9	2.2925	2935	2946	2956	2966	2976	2986	2996	3006	3016	1	2	3	4	5	6	7	8	9
10.0	2.3026																		

$$e^x$$

	0.00	0.01	0.02	0.03	0.04	0.05	0.06	0.07	0.08	0.09
0.0	1.0000	1.0101	1.0202	1.0305	1.0408	1.0513	1.0618	1.0725	1.0833	1.0942
0.1	1.1052	1.1163	1.1275	1.1388	1.1503	1.1618	1.1735	1.1853	1.1972	1.2092
0.2	1.2214	1.2337	1.2461	1.2586	1.2712	1.2840	1.2969	1.3100	1.3231	1.3364
0.3	1.3499	1.3634	1.3771	1.3910	1.4049	1.4191	1.4333	1.4477	1.4623	1.4770
0.4	1.4918	1.5068	1.5220	1.5373	1.5527	1.5683	1.5841	1.6000	1.6161	1.6323
0.5	1.6487	1.6653	1.6820	1.6989	1.7160	1.7333	1.7507	1.7683	1.7860	1.8040
0.6	1.8221	1.8404	1.8589	1.8776	1.8965	1.9155	1.9348	1.9542	1.9739	1.9937
0.7	2.0138	2.0340	2.0544	2.0751	2.0959	2.1170	2.1383	2.1598	2.1815	2.2034
0.8	2.2255	2.2479	2.2705	2.2933	2.3164	2.3396	2.3632	2.3869	2.4109	2.4351
0.9	2.4596	2.4843	2.5093	2.5345	2.5600	2.5857	2.6117	2.6379	2.6645	2.6912
1.0	2.7183	2.7456	2.7732	2.8011	2.8292	2.8577	2.8864	2.9154	2.9447	2.9743
1.1	3.0042	3.0344	3.0649	3.0957	3.1268	3.1582	3.1899	3.2220	3.2544	3.2871
1.2	3.3201	3.3535	3.3872	3.4212	3.4556	3.4903	3.5254	3.5609	3.5966	3.6328
1.3	3.6693	3.7062	3.7434	3.7810	3.8190	3.8574	3.8962	3.9354	3.9749	4.0149
1.4	4.0552	4.0960	4.1371	4.1787	4.2207	4.2631	4.3060	4.3492	4.3929	4.4371
1.5	4.4817	4.5267	4.5722	4.6182	4.6646	4.7115	4.7588	4.8066	4.8550	4.9037
1.6	4.9530	5.0028	5.0531	5.1039	5.1552	5.2070	5.2593	5.3122	5.3656	5.4195
1.7	5.4739	5.5290	5.5845	5.6407	5.6973	5.7546	5.8124	5.8709	5.9299	5.9895
1.8	6.0496	6.1104	6.1719	6.2339	6.2965	6.3598	6.4237	6.4883	6.5535	6.6194
1.9	6.6859	6.7531	6.8210	6.8895	6.9588	7.0287	7.0993	7.1707	7.2427	7.3155
2.0	7.3891	7.4633	7.5383	7.6141	7.6906	7.7679	7.8460	7.9248	8.0045	8.0849
2.1	8.1662	8.2482	8.3311	8.4149	8.4994	8.5849	8.6711	8.7583	8.8463	8.9352
2.2	9.0250	9.1157	9.2073	9.2999	9.3933	9.4877	9.5831	9.6794	9.7767	9.8749
2.3	9.9742	10.074	10.176	10.278	10.381	10.486	10.591	10.697	10.805	10.913
2.4	11.023	11.134	11.246	11.359	11.473	11.588	11.705	11.822	11.941	12.061
2.5	12.182	12.305	12.429	12.554	12.680	12.807	12.936	13.066	13.197	13.330
2.6	13.464	13.599	13.736	13.874	14.013	14.154	14.296	14.440	14.585	14.732
2.7	14.880	15.029	15.180	15.333	15.487	15.643	15.800	15.959	16.119	16.281
2.8	16.445	16.610	16.777	16.945	17.116	17.288	17.462	17.637	17.814	17.993
2.9	18.174	18.357	18.541	18.728	18.916	19.106	19.298	19.492	19.688	19.886
3.0	20.086	20.287	20.491	20.697	20.905	21.115	21.328	21.542	21.758	21.977
3.1	22.198	22.421	22.646	22.874	23.104	23.336	23.571	23.807	24.047	24.288
3.2	24.533	24.779	25.028	25.280	25.534	25.790	26.050	26.311	26.576	26.843
3.3	27.113	27.385	27.660	27.938	28.219	28.503	28.789	29.079	29.371	29.666
3.4	29.964	30.265	30.569	30.877	31.187	31.500	31.817	32.137	32.460	32.786
3.5	33.115	33.448	33.784	34.124	34.467	34.813	35.163	35.517	35.874	36.234
3.6	36.598	36.966	37.338	37.713	38.092	38.475	38.861	39.252	39.646	40.045
3.7	40.447	40.854	41.264	41.679	42.098	42.521	42.948	43.380	43.816	44.256
3.8	44.701	45.150	45.604	46.063	46.525	46.993	47.465	47.942	48.424	48.911
3.9	49.402	49.899	50.400	50.907	51.419	51.935	52.457	52.985	53.517	54.055
4.0	54.598	55.147	55.701	56.261	56.826	57.397	57.974	58.557	59.145	59.740
4.1	60.340	60.947	61.559	62.178	62.803	63.434	64.072	64.715	65.366	66.023
4.2	66.686	67.357	68.033	68.717	69.408	70.105	70.810	71.522	72.240	72.966
4.3	73.700	74.440	75.189	75.944	76.708	77.478	78.257	79.044	79.838	80.640
4.4	81.451	82.269	83.096	83.931	84.775	85.627	86.488	87.357	88.235	89.121
4.5	90.017	90.922	91.836	92.759	93.691	94.632	95.583	96.544	97.514	98.494
4.6	99.484	100.48	101.49	102.51	103.54	104.58	105.64	106.70	107.77	108.85
4.7	109.95	111.05	112.17	113.30	114.43	115.58	116.75	117.92	119.10	120.30
4.8	121.51	122.73	123.97	125.21	126.47	127.74	129.02	130.32	131.63	132.95
4.9	134.29	135.64	137.00	138.38	139.77	141.17	142.59	144.03	145.47	146.94

$$e^x$$

	0.00	0.01	0.02	0.03	0.04	0.05	**0.06**	0.07	0.08	0.09
5.0	148.41	149.90	151.41	152.93	154.47	156.02	157.59	159.17	160.77	162.39
5.1	164.02	165.67	167.34	169.02	170.72	172.43	174.16	175.91	177.68	179.47
5.2	181.27	183.09	184.93	186.79	188.67	190.57	192.48	194.42	196.37	198.34
5.3	200.34	202.35	204.38	206.44	208.51	210.61	212.72	214.86	217.02	219.20
5.4	221.41	223.63	225.88	228.15	230.44	232.76	235.10	237.46	239.85	242.26
5.5	244.69	247.15	249.64	252.14	254.68	257.24	259.82	262.43	265.07	267.74
5.6	270.43	273.14	275.89	278.66	281.46	284.29	287.15	290.03	292.95	295.89
5.7	298.87	301.87	304.90	307.97	311.06	314.19	317.35	320.54	323.76	327.01
5.8	330.30	333.62	336.97	340.36	343.78	347.23	350.72	354.25	357.81	361.41
5.9	365.04	368.71	372.41	376.15	379.93	383.75	387.61	391.51	395.44	399.41
6.0	403.43	407.48	411.58	415.72	419.89	424.11	428.38	432.68	437.03	441.42
6.1	445.86	450.34	454.86	459.44	464.05	468.72	473.43	478.19	482.99	487.85
6.2	492.75	497.70	502.70	507.76	512.86	518.01	523.22	528.48	533.79	539.15
6.3	544.57	550.04	555.57	561.16	566.80	572.49	578.25	584.06	589.93	595.86
6.4	601.85	607.89	614.00	620.17	626.41	632.70	639.06	645.48	651.97	658.52
6.5	665.14	671.83	678.58	685.40	692.29	699.24	706.27	713.37	720.54	727.78
6.6	735.10	742.48	749.95	757.48	765.09	772.78	780.55	788.40	796.32	804.32
6.7	812.41	820.57	828.82	837.15	845.56	854.06	862.64	871.31	880.07	888.91
6.8	897.85	906.87	915.99	925.19	934.49	943.88	953.37	962.95	972.63	982.40
6.9	992.27	1002.2	1012.3	1022.5	1032.8	1043.1	1053.6	1064.2	1074.9	1085.7
7.0	1096.6	1107.7	1118.8	1130.0	1141.4	1152.9	1164.4	1176.1	1188.0	1199.9
7.1	1212.0	1224.1	1236.5	1248.9	1261.4	1274.1	1286.9	1299.8	1312.9	1326.1
7.2	1339.4	1352.9	1366.5	1380.2	1394.1	1408.1	1422.3	1436.6	1451.0	1465.6
7.3	1480.3	1495.2	1510.2	1525.4	1540.7	1556.2	1571.8	1587.6	1603.6	1619.7
7.4	1636.0	1652.4	1669.0	1685.8	1702.8	1719.9	1737.1	1754.6	1772.2	1790.1
7.5	1808.0	1826.2	1844.6	1863.1	1881.8	1900.7	1919.8	1939.1	1958.6	1978.3
7.6	1998.2	2018.3	2038.6	2059.1	2079.7	2100.6	2121.8	2143.1	2164.6	2186.4
7.7	2208.3	2230.5	2253.0	2275.6	2298.5	2321.6	2344.9	2368.5	2392.3	2416.3
7.8	2440.6	2465.1	2489.9	2514.9	2540.2	2565.7	2591.5	2617.6	2643.9	2670.4
7.9	2697.3	2724.4	2751.8	2779.4	2807.4	2835.6	2864.1	2892.9	2921.9	2951.3
8.0	2981.0	3010.9	3041.2	3071.7	3102.6	3133.8	3165.3	3197.1	3229.2	3261.7
8.1	3294.5	3327.6	3361.0	3394.8	3428.9	3463.4	3498.2	3533.3	3568.9	3604.7
8.2	3641.0	3677.5	3714.5	3751.8	3789.5	3827.6	3866.1	3904.9	3944.2	3983.8
8.3	4023.9	4064.3	4105.2	4146.4	4188.1	4230.2	4272.7	4315.6	4359.0	4402.8
8.4	4447.1	4491.8	4536.9	4582.5	4628.6	4675.1	4722.1	4769.5	4817.4	4865.9
8.5	4914.8	4964.2	5014.1	5064.4	5115.3	5166.8	5218.7	5271.1	5324.1	5377.6
8.6	5431.7	5486.2	5541.4	5597.1	5653.3	5710.1	5767.5	5825.5	5884.0	5943.2
8.7	6002.9	6063.2	6124.2	6185.7	6247.9	6310.7	6374.1	6438.2	6502.9	6568.2
8.8	6634.2	6700.9	6768.3	6836.3	6905.0	6974.4	7044.5	7115.3	7186.8	7259.0
8.9	7332.0	7405.7	7480.1	7555.3	7631.2	7707.9	7785.4	7863.6	7942.6	8022.5
9.0	8103.1	8184.5	8266.8	8349.9	8433.8	8518.5	8604.2	8690.6	8778.0	8866.2
9.1	8955.3	9045.3	9136.2	9228.0	9320.8	9414.4	9509.1	9604.6	9701.2	9798.7
9.2	9897.1	9996.6	10097.	10199.	10301.	10405.	10509.	10615.	10721.	10829.
9.3	10938.	11048.	11159.	11271.	11384.	11499.	11614.	11731.	11849.	11968.
9.4	12088.	12210.	12333.	12457.	12582.	12708.	12836.	12965.	13095.	13227.
9.5	13360.	13494.	13630.	13767.	13905.	14045.	14186.	14328.	14472.	14618.
9.6	14765.	14913.	15063.	15214.	15367.	15522.	15678.	15835.	15994.	16155.
9.7	16318.	16482.	16647.	16815.	16984.	17154.	17327.	17501.	17677.	17854.
9.8	18034.	18215.	18398.	18583.	18770.	18958.	19149.	19341.	19536.	19732.
9.9	19930.	20131.	20333.	20537.	20744.	20952.	21163.	21375.	21590.	21807.

SINES OF ANGLES

°	0′	6′	12′	18′	24′	30′	36′	42′	48′	54′	1′	2′	3′	4′	5′
0	.0000	0017	0035	0052	0070	0087	0105	0122	0140	0157	3	6	9	12	15
1	.0175	0192	0209	0227	0244	0262	0279	0297	0314	0332	3	6	9	12	15
2	.0349	0366	0384	0401	0419	0436	0454	0471	0488	0506	3	6	9	12	15
3	.0523	0541	0558	0576	0593	0610	0628	0645	0663	0680	3	6	9	12	15
4	.0698	0715	0732	0750	0767	0785	0802	0819	0837	0854	3	6	9	12	14
5	.0872	0889	0906	0924	0941	0958	0976	0993	1011	1028	3	6	9	12	14
6	.1045	1063	1080	1097	1115	1132	1149	1167	1184	1201	3	6	9	12	14
7	.1219	1236	1253	1271	1288	1305	1323	1340	1357	1374	3	6	9	12	14
8	.1392	1409	1426	1444	1461	1478	1495	1513	1530	1547	3	6	9	12	14
9	.1564	1582	1599	1616	1633	1650	1668	1685	1702	1719	3	6	9	11	14
10	.1736	1754	1771	1788	1805	1822	1840	1857	1874	1891	3	6	9	11	14
11	.1908	1925	1942	1959	1977	1994	2011	2028	2045	2062	3	6	9	11	14
12	.2079	2096	2113	2130	2147	2164	2181	2198	2215	2233	3	6	9	11	14
13	.2250	2267	2284	2300	2317	2334	2351	2368	2385	2402	3	6	8	11	14
14	.2419	2436	2453	2470	2487	2504	2521	2538	2554	2571	3	6	8	11	14
15	.2588	2605	2622	2629	2656	2672	2689	2706	2723	2740	3	6	8	11	14
16	.2756	2773	2790	2807	2823	2840	2857	2874	2890	2907	3	6	8	11	14
17	.2924	2940	2957	2974	2990	3007	3024	3040	3057	3074	3	6	8	11	14
18	.3090	3107	3123	3140	3156	3173	3190	3206	3223	3239	3	6	8	11	14
19	.3256	3272	3289	3305	3322	3338	3355	3371	3387	3404	3	5	8	11	14
20	.3420	3437	3453	3469	3486	3502	3518	3535	3551	3567	3	5	8	11	14
21	.3584	3600	3616	3633	3649	3665	3681	3697	3714	3730	3	5	8	11	14
22	.3746	3762	3778	3795	3811	3827	3843	3859	3875	3891	3	5	8	11	13
23	.3907	3923	3939	3955	3971	3987	4003	4019	4035	4051	3	5	8	11	13
24	.4067	4083	4099	4115	4131	4147	4163	4179	4195	4210	3	5	8	11	13
25	.4226	4242	4258	4274	4289	4305	4321	4337	4352	4368	3	5	8	11	13
26	.4384	4399	4415	4431	4446	4462	4478	4493	4509	4524	3	5	8	10	13
27	.4540	4555	4571	4586	4602	4617	4633	4648	4664	4679	3	5	8	10	13
28	.4695	4710	4726	4741	4756	4772	4787	4802	4818	4833	3	5	8	10	13
29	.4848	4863	4879	4894	4909	4924	4939	4955	4970	4985	3	5	8	10	13
30	.5000	5015	5030	5045	5060	5075	5090	5105	5120	5135	3	5	8	10	13
31	.5150	5165	5180	5195	5210	5225	5240	5255	5270	5284	2	5	7	10	12
32	.5299	5314	5329	5344	5358	5373	5388	5402	5417	5432	2	5	7	10	12
33	.5446	5461	5476	5490	5505	5519	5534	5548	5563	5577	2	5	7	10	12
34	.5592	5606	5621	5635	5650	5664	5678	5693	5707	5721	2	5	7	10	12
35	.5736	5750	5764	5779	5793	5807	5821	5835	5850	5864	2	5	7	10	12
36	.5878	5892	5906	5920	5934	5948	5962	5976	5990	6004	2	5	7	9	12
37	.6018	6032	6046	6060	6074	6088	6101	6115	6129	6143	2	5	7	9	12
38	.6157	6170	6184	6198	6211	6225	6239	6252	6266	6280	2	5	7	9	11
39	.6293	6307	6320	6334	6347	6361	6374	6388	6401	6414	2	4	7	9	11
40	.6428	6441	6455	6468	6481	6494	6508	6521	6534	6547	2	4	7	9	11
41	.6561	6574	6587	6600	6613	6626	6639	6652	6665	6678	2	4	7	9	11
42	.6691	6704	6717	6730	6743	6756	6769	6782	6794	6807	2	4	6	9	11
43	.6820	6833	6845	6858	6871	6884	6896	6909	6921	6934	2	4	6	8	11
44	.6947	6959	6972	6984	6997	7009	7022	7034	7046	7059	2	4	6	8	10

SINES OF ANGLES

°	0'	6'	12'	18'	24'	30'	36'	42'	48'	54'	1'	2'	3'	4'	5'
45	.7071	7083	7096	7108	7120	7133	7145	7157	7169	7181	2	4	6	8	10
46	.7193	7206	7218	7230	7242	7254	6266	7278	7290	7302	2	4	6	8	10
47	.7314	7325	7337	7349	7361	7373	7385	7396	7408	7420	2	4	6	8	10
48	.7431	7443	7455	7466	7478	7490	7501	7513	7524	7536	2	4	6	8	10
49	.7547	7559	7570	7581	7593	7604	7615	7627	7638	7649	2	4	6	8	9
50	.7660	7672	7683	7694	7705	7716	7727	7738	7749	7760	2	4	6	7	9
51	.7771	7782	7793	7804	7815	7826	7837	7848	7859	7869	2	4	5	7	9
52	.7880	7891	7902	7912	7923	7934	7944	7955	7965	7976	2	4	5	7	9
53	.7986	7997	8007	8018	8028	8039	8049	8059	8070	8080	2	3	5	7	9
54	.8090	8100	8111	8121	8131	8141	8151	8161	8171	8181	2	3	5	7	8
55	.8192	8202	8211	8221	8231	8241	8251	8261	8271	8281	2	3	5	7	8
56	.8290	8300	8310	8320	8329	8339	8348	8358	8368	8377	2	3	5	6	8
57	.8387	8396	8406	8415	8425	8434	8443	8453	8462	8471	2	3	5	6	8
58	.8480	8490	8499	8508	8517	8526	8536	8545	8554	8563	2	3	5	6	8
59	.8572	8581	8590	8599	8607	8616	8625	8634	8643	8652	1	3	4	6	7
60	.8660	8669	8678	8686	8695	8704	8712	8721	8729	8738	1	3	4	6	7
61	.8746	8755	8763	8771	8780	8788	8796	8805	8813	8821	1	3	4	6	7
62	.8829	8838	8846	8854	8862	8870	8878	8886	8894	8902	1	3	4	5	7
63	.8910	8918	8926	8934	8942	8949	8957	8965	8973	8980	1	3	4	5	6
64	.8988	8996	9003	9011	9018	9026	9033	9041	9048	9056	1	3	4	5	6
65	.9063	9070	9078	9085	9092	9100	9107	9114	9121	9128	1	2	4	5	6
66	.9135	9143	9150	9157	9164	9171	9178	9184	9191	9198	1	2	3	5	6
67	.9205	9212	9219	9225	9232	9239	9245	9252	9259	9265	1	2	3	4	6
68	.9272	9278	9285	9291	9298	9304	9311	9317	9323	9330	1	2	3	4	5
69	.9336	9342	9348	9354	9361	9367	9373	9379	9385	9391	1	2	3	4	5
70	.9297	9403	9409	9415	9421	9426	9432	9438	9444	9449	1	2	3	4	5
71	.9455	9461	9466	9472	9478	9483	9489	9494	9500	9505	1	2	3	4	5
72	.9511	9516	9521	9527	9532	9537	9542	9548	9553	9558	1	2	3	4	4
73	.9563	9568	9573	9578	9583	9588	9593	9598	9603	9608	1	2	2	3	4
74	.9613	9617	9622	9627	9632	9636	9641	9646	9650	9655	1	2	2	3	4
75	.9659	9664	9668	9673	9677	9681	9686	9690	9694	9699	1	1	2	3	4
76	.9703	9707	9711	9715	9720	9724	9728	9732	9736	9740	1	1	2	3	3
77	.9744	9748	9751	9755	9759	9763	9767	9770	9774	9778	1	1	2	3	3
78	.9781	9785	9789	9792	9796	9799	9803	9806	9810	9813	1	1	2	2	3
79	.9816	9820	9823	9826	9829	9833	9836	9839	9842	9845	1	1	2	2	3
80	.9848	9851	9854	9857	9860	9863	9866	9869	9871	9874	0	1	1	2	2
81	.9877	9880	9882	9885	9888	9890	9893	9895	9898	9900	0	1	1	2	2
82	.9903	9905	9907	9910	9912	9914	9917	9919	9921	9923	0	1	1	2	2
83	.9925	9928	9930	9932	9934	9936	9938	9940	9942	9943	0	1	1	1	2
84	.9945	9947	9949	9951	9952	9954	9956	9957	9959	9960	0	1	1	1	1
85	.9962	9963	9965	9966	9968	9969	9971	9972	9973	9974	0	0	1	1	1
86	.9976	9977	9978	9979	9980	9981	9982	9983	9984	9985	0	0	1	1	1
87	.9986	9987	9988	9989	9990	9990	9991	9992	9993	9993	0	0	0	0	0
88	.9994	9995	9995	9996	9996	9997	9997	9997	9998	9998	0	0	0	0	0
89	.9998	9999	9999	9999	9999	1.000	1.000	1.000	1.000	1.000	0	0	0	0	0

COSINES OF ANGLES

°	0'	6'	12'	18'	24'	30'	36'	42'	48'	54'	SUBTRACT				
											1'	2'	3'	4	5
0	1.000	1.000	1.000	1.000	1.000	1.000	**9999**	**9999**	**9999**	**9999**	0	0	0	0	0
1	.9998	9998	9998	9997	9997	9997	9996	9996	9995	9995	0	0	0	0	0
2	.9994	9993	9993	9992	9991	9990	9990	9989	9988	9987	0	0	0	0	0
3	.9986	9985	9984	9983	9982	9981	9980	9979	9978	9977	0	0	1	1	1
4	.9976	9974	9973	9972	9971	9969	9968	9966	9965	9963	0	0	1	1	1
5	.9962	9960	9959	9957	9956	9954	9952	9951	9949	9947	0	1	1	1	1
6	.9945	9943	9942	9940	9938	9936	9934	9932	9930	9928	0	1	1	1	2
7	.9925	9923	9921	9919	9917	9914	9912	9910	9907	9905	0	1	1	2	2
8	.9903	9900	9898	9895	9893	9890	9888	9885	9882	9880	0	1	1	2	2
9	.9877	9874	9871	9869	9866	9863	9860	9857	9854	9851	0	1	1	2	2
10	.9848	9845	9842	9839	9836	9833	9829	9826	9823	9820	1	1	2	2	3
11	.9816	9813	9810	9806	9803	9799	9796	9792	9789	9785	1	1	2	2	3
12	9781	9778	9774	9770	9767	9763	9759	9755	9751	9748	1	1	2	3	3
13	.9744	9740	9736	9732	9728	9724	9720	9715	9711	9707	1	1	2	3	3
14	.9703	9699	9694	9690	9686	9681	9677	9673	9668	9664	1	1	2	3	4
15	.9659	9655	9650	9646	9641	9636	9632	9627	9622	9617	1	2	2	3	4
16	.9613	9608	9603	9598	9593	9588	9583	9578	9573	9568	1	2	2	3	4
17	.9563	9558	9553	9548	9542	9537	9532	9527	9521	9516	1	2	3	4	4
18	.9511	9505	9500	9494	9489	9483	9478	9472	9466	9461	1	2	3	4	5
19	.9455	9449	9444	9438	9432	9426	9421	9415	9409	9403	1	2	3	4	5
20	.8397	9391	9385	9379	9373	9367	9361	9354	9348	9342	1	2	3	4	5
21	.9336	9330	9323	9317	9311	9304	9298	9291	9285	9278	1	2	3	4	5
22	.9272	9265	9259	9252	9245	9239	9232	9225	9219	9212	1	2	3	4	6
23	.9205	9198	9191	9184	9178	9171	9164	9157	9150	9143	1	2	3	5	6
24	.9135	9128	9121	9114	9107	9100	9092	9085	9078	9070	1	2	4	5	6
25	.9063	9056	9048	9041	9033	9026	9018	9011	9003	8996	1	3	4	5	6
26	.8988	8980	8973	8965	8957	8949	8942	8934	8926	8918	1	3	4	5	6
27	.8910	8902	8894	8886	8878	8870	8862	8854	8846	8838	1	3	4	5	7
28	.8829	8821	8813	8805	8796	8788	8780	8771	8763	8755	1	3	4	6	7
29	.8746	8738	8729	8721	8712	8704	8695	8686	8678	8669	1	3	4	6	7
30	.8660	8652	8643	8634	8625	8616	8607	**9599**	**9590**	8581	1	3	4	6	7
31	.8572	8563	8554	8545	8536	8526	8517	8508	8499	8490	2	3	5	6	8
32	.8480	8471	8462	8453	8443	8434	8425	8415	8406	8396	2	3	5	6	8
33	.8387	8377	8368	8358	8348	8339	8329	8320	8310	8300	2	3	5	6	8
34	.8290	8281	8271	8261	8251	8241	8231	8221	8211	8202	2	3	5	7	8
35	.8192	8181	8171	8161	8151	8141	8131	8121	8111	8100	2	3	5	7	8
36	.8090	8080	8070	8059	8049	8039	8028	8018	8007	7997	2	3	5	7	9
37	.7986	7976	7965	7955	7944	7934	7923	7912	7902	7891	2	4	5	7	9
38	.7880	7869	7859	7848	7837	7826	7815	7804	7793	7782	2	4	5	7	9
39	.7771	7760	7749	7738	7727	7716	7705	7694	7683	7672	2	4	6	7	9
40	.7660	7649	7638	7627	7615	7604	7593	7581	7570	7559	2	4	6	8	9
41	.7547	7536	7524	7513	7501	7490	7478	7466	7455	7443	2	4	6	8	10
42	.7431	7420	7408	7396	7385	7373	7361	7349	7337	7325	2	4	6	8	10
43	.7314	7302	7290	7278	7266	7254	7242	7230	7218	7206	2	4	6	8	10
44	.7193	7181	7169	7157	7145	7133	7120	7108	7096	7083	2	4	6	8	10

Figures in bold type show change of integer   SUBTRACT

COSINES OF ANGLES

°	0′	6′	12′	18′	24′	30′	36′	42′	48′	54′	1′ 2′ 3′ 4′ 5′
											SUBTRACT
45	.7071	7059	7046	7034	7022	7009	6997	6984	6972	6959	2 4 6 8 10
46	.6947	6934	6921	6909	6896	6884	6871	6858	6845	6833	2 4 6 8 11
47	.6820	6807	6794	6782	6769	6756	6743	6730	6717	6704	2 4 6 9 11
48	.6691	6678	6665	6652	6639	6626	6613	6600	6587	6574	2 4 7 9 11
49	.6561	6547	6534	6521	6508	6494	6481	6468	6455	6441	2 4 7 9 11
50	.6428	6414	6401	6388	6374	6361	6347	6334	6320	6307	2 4 7 9 11
51	.6293	6280	6266	6252	6239	6225	6211	6198	6184	6170	2 5 7 9 11
52	.6157	6143	6129	6115	6101	6088	6074	6060	6046	6032	2 5 7 9 12
53	.6018	6004	5990	5976	5962	5948	5934	5920	5906	5892	2 5 7 9 12
54	.5878	5864	5850	5825	5821	5807	5793	5779	5764	5750	2 5 7 9 12
55	.5736	5721	5707	5693	5678	5664	5650	5635	5621	5606	2 5 7 10 12
56	.5592	5577	5563	5548	5534	5519	5505	5490	5476	5461	2 5 7 10 12
57	.5446	5432	5417	5402	5388	5373	5358	5344	5329	5314	2 5 7 10 12
58	.5299	5284	5270	5255	5240	5225	5210	5195	5180	5165	2 5 7 10 12
59	.5150	5135	5120	5105	5090	5075	5060	5045	5030	5015	3 5 8 10 13
60	.5000	4985	4970	4955	4909	4924	4909	4894	4879	4863	3 5 8 10 13
61	.4848	4833	4818	4802	4787	4772	4756	4741	4726	4710	3 5 8 10 13
62	.4695	4679	4664	4648	4633	4617	4602	4586	4571	4555	3 5 8 10 13
63	.4540	4524	4509	4493	4478	4462	4446	4431	4415	4399	3 5 8 10 13
64	.4384	4368	4352	4337	4321	4305	4289	4274	4258	4242	3 5 8 11 13
65	.4226	4210	4195	4179	4163	4147	4131	4115	4099	4083	3 5 8 11 13
66	.4067	4051	4035	4019	4003	3987	3971	3955	3939	3923	3 5 8 11 13
67	.3907	3891	3875	3859	3843	3827	3811	3795	3778	3762	3 5 8 11 13
68	.3746	3730	3714	3697	3681	3665	3649	3633	3616	3600	3 5 8 11 14
69	.3584	3567	3551	3535	3518	3502	3486	3469	3453	3437	3 5 8 11 14
70	.3420	3404	3387	3371	3355	3338	3322	3305	3289	3272	3 5 8 11 14
71	.3256	3239	3223	3206	3190	3173	3156	3140	3123	3107	3 6 8 11 14
72	.3090	3074	3057	3040	3024	3007	2990	2974	2957	2940	3 6 8 11 14
73	.2924	2907	2890	2874	2857	2840	2823	2807	2790	2773	3 6 8 11 14
74	.2756	2740	2723	2706	2689	2672	2656	2639	2622	2605	3 6 8 11 14
75	.2588	2571	2554	2538	2521	2504	2487	2470	2453	2436	3 6 8 11 14
76	.2419	2402	2385	2368	2351	2334	2317	2300	2284	2267	3 6 8 11 14
77	.2250	2233	2215	2198	2181	2164	2147	2130	2113	2096	3 6 9 11 14
78	.2079	2062	2045	2028	2011	1994	1977	1959	1942	1925	3 6 9 11 14
79	.1908	1891	1874	1857	1840	1822	1805	1788	1771	1754	3 6 9 11 14
80	.1736	1719	1702	1685	1668	1650	1633	1616	1599	1582	3 6 9 11 14
81	.1564	1547	1530	1513	1495	1478	1461	1444	1426	1409	3 6 9 12 14
82	.1392	1374	1357	1340	1323	1305	1288	1271	1253	1236	3 6 9 12 14
83	.1219	1201	1184	1167	1149	1132	1115	1097	1080	1063	3 6 9 12 14
84	.1045	1028	1011	0993	0976	0958	0941	0924	0906	0889	3 6 9 12 14
85	.0872	0854	0837	0819	0802	0785	0767	0750	0732	0715	3 6 9 12 14
86	.0698	0680	0663	0645	0628	0610	0593	0576	0558	0541	3 6 9 12 15
87	.0523	0506	0488	0471	0454	0436	0419	0401	0384	0366	3 6 9 12 15
88	.0349	0332	0314	0297	0279	0262	0244	0227	0209	0192	3 6 9 12 15
89	.0175	0157	0140	0122	0105	0087	0070	0052	0035	0017	3 6 9 12 15

SUBTRACT

TANGENTS OF ANGLES

°	0′	6′	12′	18′	24′	30′	36′	42′	48′	54′	1′	2′	3′	4′	5′
0	.0000	0017	0035	0052	0070	0087	0105	0122	1040	0157	3	6	9	12	15
1	.0175	0192	0209	0227	0244	0262	0279	0297	0314	0332	3	6	9	12	15
2	.0349	0367	0384	0402	0419	0437	0454	0472	0489	0507	3	6	9	12	15
3	.0524	0542	0559	0577	0594	0612	0629	0647	0664	0682	3	6	9	12	15
4	.0699	0717	0734	0752	0769	0787	0805	0822	0840	0857	3	6	9	12	15
5	.0875	0892	0910	0928	0945	0963	0981	0998	1016	1033	3	6	9	12	15
6	.1051	1069	1086	1104	1122	1139	1157	1175	1192	1210	3	6	9	12	15
7	.1228	1246	1263	1281	1299	1317	1334	1352	1370	1388	3	6	9	12	15
8	.1405	1423	1441	1459	1477	1495	1512	1530	1548	1566	3	6	9	12	15
9	.1584	1602	1620	1638	1655	1673	1691	1709	1727	1745	3	6	9	12	15
10	.1763	1781	1799	1817	1835	1853	1871	1890	1908	1926	3	6	9	12	15
11	.1944	1962	1980	1998	2016	2035	2053	2071	2089	2107	3	6	9	12	15
12	.2126	2144	2162	2180	2199	2217	2235	2254	2272	2290	3	6	9	12	15
13	.2309	2327	2345	2364	2382	2401	2419	2438	2456	2475	3	6	9	12	15
14	.2493	2512	2530	2549	2568	2586	2605	2623	2642	2661	3	6	9	12	16
15	.2679	2698	2717	2736	2754	2773	2792	2811	2830	2849	3	6	9	13	16
16	.2867	2886	2905	2924	2943	2962	2981	3000	3019	3038	3	6	9	13	16
17	.3057	3076	3096	3115	3134	3153	3172	3191	3211	3230	3	6	10	13	16
18	.3249	3269	3288	3307	3327	3346	3365	3385	3404	3424	3	6	10	13	16
19	.3443	3463	3482	3502	3522	3541	3561	3581	3600	3620	3	7	10	13	16
20	.3640	3659	3679	3699	3719	3739	3759	3779	3799	3819	3	7	10	13	17
21	.3839	3859	3879	3899	3919	3939	3959	3979	4000	4020	3	7	10	13	17
22	.4040	4061	4081	4101	4122	4142	4163	4183	4204	4224	3	7	10	14	17
23	.4245	4265	4286	4307	4327	4348	4369	4390	4411	4431	3	7	10	14	17
24	.4452	4473	4494	4515	4536	4557	4578	4599	4621	4642	4	7	11	14	18
25	.4663	4684	4706	4727	4748	4770	4791	4813	4834	4856	4	7	11	14	18
26	.4877	4899	4921	4942	4964	4986	5008	5029	5051	5073	4	7	11	15	18
27	.5095	5117	5139	5161	5184	5206	5228	5250	5272	5295	4	7	11	15	18
28	.5317	5340	5362	5384	5407	5430	5452	5475	5498	5520	4	8	11	15	19
29	.5543	5566	5589	5612	5635	5658	5681	5704	5727	5750	4	8	12	15	19
30	.5774	5797	5820	5844	5867	5890	5914	5938	5961	5985	4	8	12	16	20
31	.6009	6032	6056	6080	6104	6128	6152	6176	6200	6224	4	8	12	16	20
32	.6249	6273	6297	6322	6346	6371	6395	6420	6445	6469	4	8	12	16	20
33	.6494	6519	6544	6569	6594	6619	6644	6669	6694	6720	4	8	13	17	21
34	.6745	6771	6796	6822	6847	6873	6899	6924	6950	6976	4	9	13	17	21
35	.7002	7028	7054	7080	7107	7133	7159	7186	7212	7239	4	9	13	18	22
36	.7265	7292	7319	7346	7373	7400	7427	7454	7481	7508	5	9	14	18	23
37	.7536	7563	7590	7618	7646	7673	7701	7729	7757	7785	5	9	14	18	23
38	.7813	7841	7869	7898	7926	7954	7983	8012	8040	8069	5	9	14	19	24
39	.8098	8127	8156	8185	8214	8243	8273	8302	8332	8361	5	10	15	20	24
40	.8391	8421	8451	8481	8511	8541	8571	8601	8632	8662	5	10	15	20	25
41	.8693	8724	8754	8785	8816	8847	8878	8910	8941	8972	5	10	16	21	26
42	.9004	9036	9067	9099	9131	9163	9195	9228	9260	9293	5	11	16	21	27
43	.9325	9358	9391	9424	9457	9490	9523	9556	9590	9623	6	11	17	22	28
44	.9657	9691	9725	9759	9793	9827	9861	9896	9930	9965	6	11	17	23	29

TANGENTS OF ANGLES

°	0′	6′	12′	18′	24′	30′	36′	42′	48′	54′	1′ 2′ 3′ 4′ 5′
45	1.0000	0035	0070	0105	0141	0176	0212	0247	0283	0319	6 12 18 24 30
46	1.0355	0392	0428	0464	0501	0538	0575	0612	0649	0686	6 12 18 25 31
47	1.0724	0761	0799	0837	0875	0913	0951	0990	1028	1067	6 13 19 25 32
48	1.1106	1145	1184	1224	1263	1303	1343	1383	1423	1463	7 13 20 26 33
49	1.1504	1544	1585	1626	1667	1708	1750	1792	1833	1875	7 14 21 28 34
50	1.1918	1960	2002	2045	2088	2131	2174	2218	2261	2305	7 14 22 29 36
51	1.2349	2393	2437	2482	2527	2572	2617	2662	2708	2753	8 15 23 30 38
52	1.2799	2846	2892	2938	2985	3032	3079	3127	3175	3222	8 16 24 31 39
53	1.3270	3319	3367	3416	3465	3514	3564	3613	3663	3713	8 16 25 33 41
54	1.3764	3814	3865	3916	3968	4019	4071	4124	4176	4229	9 17 26 34 43
55	1.4281	4335	4388	4442	4496	4550	4605	4659	4715	4770	9 18 27 36 45
56	1.4826	4882	4938	4994	5051	5108	5166	5224	5282	5340	10 19 29 38 48
57	1.5399	5458	5517	5577	5637	5697	5757	5818	5880	5941	10 20 30 40 50
58	1.6003	6066	6128	6191	6255	6319	6383	6447	6512	6577	11 21 32 43 53
59	1.6643	6709	6775	6842	6909	6977	7045	7113	7182	7251	11 23 34 45 56
60	1.7321	7391	7461	7532	7603	7675	7747	7820	7893	7966	12 24 36 48 60
61	1.8040	8115	8190	8265	8341	8418	8495	8572	8650	8728	13 26 38 51 64
62	1.8807	8887	8967	9047	9128	9210	9292	9375	9458	9542	14 27 41 55 68
63	1.9626	9711	9797	9883	9970	**0057**	**0145**	**0233**	**0323**	**0413**	15 29 44 58 73
64	2.0503	0594	0686	0778	0872	0965	1060	1155	1251	1348	16 31 47 63 78
65	2.1445	1543	1642	1742	1842	1943	2045	2148	2251	2355	17 34 51 68 85
66	2.2460	2566	2673	2781	2889	2998	3109	3220	3332	3445	18 37 55 73 92
67	2.3559	3673	3789	3906	4023	4142	4262	4383	4504	4627	20 40 60 79 99
68	2.4751	4876	5002	5129	5257	5386	5517	5649	5782	5916	22 43 65 87 108
69	2.6051	6187	6325	6464	6605	6746	6889	7034	7179	7326	24 47 71 95 118
70	2.7475	7625	7776	7929	8083	8239	8397	8556	8716	8878	26 52 78 104 130
71	2.9042	9208	9375	9544	9714	9887	**0061**	**0237**	**0415**	**0595**	29 58 87 116 145
72	3.0777	0961	1146	1334	1524	1716	1910	2106	2305	2506	32 64 96 129 161
73	3.2709	2914	3122	3332	3544	3759	3977	4197	4420	4646	36 72 108 144 180
74	3.4874	5105	5339	5576	5816	6059	6305	6554	6806	7062	41 81 122 163 203
75	3.7321	7583	7848	8118	8391	8667	8947	9232	9520	9812	
76	4.0108	0408	0713	1022	1335	1653	1976	2303	2635	2972	
77	4.3315	3662	4015	4373	4737	5107	5483	5864	6252	6646	
78	4.7046	7453	7867	8288	8716	9152	9594	**0045**	**0504**	**0970**	
79	5.1446	1929	2422	2924	3435	3955	4486	5026	5578	6140	
80	5.6713	7297	7894	8502	9124	9758	**0405**	**1066**	**1742**	**2432**	
81	6.3138	3859	4596	5350	6122	6912	7720	8548	9395	**0264**	
82	7.1154	2066	3002	3962	4947	5958	6996	8062	9158	**0285**	
83	8.1443	2636	3863	5126	6427	7769	9152	**0579**	**2052**	**3572**	
84	9.5144	6768	8448	10.02	10.20	10.39	10.58	10.78	10.99	11.20	
85	11.43	11.66	11.91	12.16	12.43	12.71	13.00	13.30	13.62	13.95	
86	14.30	14.67	15.06	15.46	15.89	16.35	16.83	17.34	17.89	18.46	
87	19.08	19.74	20.45	21.20	22.02	22.90	23.86	24.90	26.03	27.27	
88	28.64	30.14	31.82	33.69	35.80	38.19	40.92	44.07	47.74	52.08	
89	57.29	63.66	71.62	81.85	95.49	114.6	143.2	191.0	286.5	573.0	

Figures in bold type show changes of integer

Index

Bold numbers indicate a main section, which may extend over more than one page, and/or a definition. *Italicised* numbers refer to pages containing relevant text figures.